AF552499

TRACE ANALYSIS OF URANIUM AND THORIUM

TRACE ANALYSIS OF URANIUM AND THORIUM

DR. TOMMANDRU AJARAIAH SAMUEL DEVRAJ
M. Sc., Ph.D.
Lecturer in Chemistry
Andhra Christian College
Guntur - 522003, Andhra Pradesh, India

Editor
Dr. Digumarti Bhaskara Rao
M.Sc., M.A., M.A., M.Ed., Ph.D.

Correspondent
G.N.R. & U.R. Noble Residential Junior College
Pattabhipuram - Shyamalanagr
Guntur - 522 006

1997
DISCOVERY PUBLISHING HOUSE
NEW DELHI - 110002

First Published - 1997

ISBN 81-7141-375-7

Published by :

Discovery Publishing House
4831/24, Ansari Road, Darya Ganj,
New Delhi - 110 002 (INDIA)
Phone : 327 9245
Fax. : 91-11-3253475

Laser Typeset by :

Debug Computer Services
Delhi.

Printed at :

Efficient Offset Printers, Delhi--110 035

CONTENTS

ACKNOWLEDGEMENTS

It is highly gratifying to realise that the Lord Almighty has been with me in presenting this book.

I express my deep sense of gratitude to Prof. Y. Anjaneyulu, Head, Centre for Environment, Jawaharlal Nehru Technological University, Hyderabad for the inspiration he created in me.

I am thankful to Mr. Ganta Nageshwara Rao and Mrs. Ganta Usha Rani, G.N.R. & U.R. Noble Residential Junior College, Guntur who paved the way for the publication of this book.

I am grateful to Dr. Digumarti Bhaskara Rao for editing this book.

It is fitting to note that the support given by my wife and children in completing the work.

Dr. T.A.S. Devraj

PREFACE

Research on new methods for trace metal analysis has grown multidimensionally in the last two decades with the introduction of several new metullurgical products of specific precise compositions for certain technical requirements. Effective analytical control at different stages of production of high quality materials sometimes involving minute quantities or concentration of analyte requires rapid and continuous search for new methods in trace analysis. Though many instrumental methods like atomic absorption spectrometry (AAS), X-ray Fluorescence (XRF) Polarography, neutron activation analysis (NAA), etc., are used in trace metal analysis, Extraction spectrophotometric Methods occupy special position due to their ease in application, simplicity, cost factor of the instrument and at the same time high sensitivity and selectivity with proper manipulation of the experimental conditions. Spectrophotometry in combination with solvent extraction plays double role like increasing the relative concentration of analyte, and separation of the analyte from the matrix components which frequently interfere with the analyte determination in many methods, and thus will be a versatile technique.

The analytical chemistry of uranium and thorium is receiving much attention in recent times with expanding nuclear reactor programmes and development of their new alloys with specific technological properties. In view of the poor sensitivity and complicated spectral interference, Atomic Absorption and Atomic Emission (ICP) methods have a low preference in the analytical chemistry or Uranium and thorium when compared with spectrophotometric methods.

Extractive spectrophotometric methods play a very important role in trace analysis of uranium and thorium complex materials as they can be made very selective and sensitive by properly modifying and optimising experimental conditions such that they are suitable for the analysis of a particular analyte in a specific matrix.

Pyridylazo compounds with a hydroxy group in the ortho position with respect to the azo group are valuable chromogenic reagents. They have been widely used in complexometric and extraction photometric methods for the determination of many ions. Highly sensitive regents have been obtained by the introduction of the diethylamino group, which is a strongly electron-donating group in the para position next to the azo group. The azo dye 2 - (2 - Pyridyl azo) - 5 - diethyl amino phenol (PADAP) and its bromine substituted derivatives 2 - (3,5 - dibromo - 2 - pyridylazo) - 5 - diethyl amino phenol (3,5 di Br PADAP), the latter having a greater tolerance to iron (iii) offer a sensitive reagents for spectro photometric determination of uranium and thorium.

The 2 - pyridylazo compounds are selective neither as chromogenic agents nor as extractants, though they give intensely coloured reactions. However, some measures can be taken to improve the situation. The most important ones are adjustment of the pH and the use of masking agents. Hydrogen ion competes with the metal ion for combination with the 2 - pyridylazo compound. Consequently, the higher the stability of the metal complex, the lower the pH at which it can exist. Therefore, at lower pH only a few number of metals will be complexed. The pH control in combination with masking agents will be useful both for selective colour formation and extraction.

The use of pyridylazo compounds as chromogenic agents can be frequently combined with solvent extraction techniques by the following methods.

1. The azo reagent is added to the aqueous solution and the chelate formed is extracted into an organic phase and is then subjected to spectro photometric measurements.
2. The metal is first extracted from the aqueous solution into an organic phase by means of a suitable organic extractant, and the azo reagent is then added to the organic phase. The colour developed is evaluated photometrically.

The addition of a surfactant to metal-dye complex, may lead to red shifts in absorption bands and increases its molar absorptivity. Usually the metal-chelate complexes formed in micellar system are more stable than those formed in the absence of micelles. Micelles are responsible for many practical applications in spectrophotometric analysis owing to their effect on the coloured complexes like :

(1) altering the micro-environment

(2) Catalysis by concentration effect in the micellar pseudo phase, and

(3) alteration of reaction path-ways, rates and equilibria with optically transparent and stable systems.

Work has been carried out on the development of rapid and

selective procedures for (a) the bulk separation of matrix and constituent components from uranium and thorium when they are present at trace level in various diversified complex, natural and synthetic industrial materials like ores, soil samples, waste waters, alloys, etc. and (b) precise and sensitive extraction spectrophotometric procedure for the trace determination of uranium and thorium.

The entire work is organized into two sections constituting 9 chapters. In section A, the investigations on uranium are described. In chapter 1, the importance of various natural and synthetic materials with varying levels of uranium and various practised methods for separation and estimation of uranium are reviewed.

Systematic and detailed investigations have been made to develop a highly selective separation of uranium (VI) by extraction with dibenzoyl methane into 15% pyridine + 10% n-butanol in chloroform mixed solvent and then followed by ternary complex formation in chloroform.

Unlike other metals, uranium (VI) forms weak complexes with fluoride, EDTA or CYDTA which can be used as masking agents for decreasing the interference of co-extracting metal ions. If EDTA is used along with magnesium chloride it enhances the extraction kinctics of Uranium (VI) facilitating for a fast selective separation of uranium from co-existing matrix components.

As EDTA and fluoride form very weak auxiliary complexes with uranium (VI) when compared with dibenzoylmethane they can be tolerated in large concentration in the extraction of uranium with dibenzoylmethane in the pH range 4-7. Further they prevent the hydrolysis of uranium and mask the interference of many other metal ions which form strong colourless complexes with EDTA and fluoride. Further, any excess of these masking agents can be removed by the addition of aluminium nitrate and calcium nitrate as salting-out agents.

The sulphate ion in amounts upto 2 mg/ml has no appreciable effect on the completeness of the extraction of uranium. If it is present in large amounts its interference can be removed by using calcium nitrate as the salting-out agent. If phosphates and sulphates are present simultaneously, a salting -out agent which consists of a mixture of aluminium and calcium or iron and calcium nitrate can suppress their interference upto 0.5 mg.

The interference of other anions and cations on extraction of U(VI) - DBM - pyridine system has been studied in presence of EDTA and sodium fluoride as masking agents and calcium nitrate and aluminium nitrate as salting-out agents. Cd, Zn, Cr, Mg, Ba and Sb do not interfere even in the absence of masking mixture while the masking mixture of EDTA and sodium fluoride binds Hg, Pb, Cu, Sn, Pt, Fe, Ni, Co, Mn, Nb, Zn and remove their interference in the extraction of uranium (VI) with U(VI) - DBM - pyridine system. However, serious interference is caused by W, Th, arsenate, carbonate, phosphate, tartrate and citrate.

Extraction of uranium (VI) - DBM system in presence of EDTA and sodium fluoride as masking agents and calcium nitrate and aluminium nitrate as salting-out agents were critically examined and optimum conditions for selective extraction of uranium are established. The data obtained from the above discussed solvent extraction studies on U(VI) are presented in chapter 2.

In chapter 3 the detailed investigation on the extractive spectrophotometric system of U(VI) - DBM - 3,5 di Br PADAP are presented. The various spectrophotometric methods of uranium and principles and precautions of spectrophotometry are briefly reviewed.

The selectivity extracted colourless uranium (VI) - DBM complex can be converted into an intensely coloured mixed ligand complex by reacting with a chelating chromogenic complexing agent like 3,5 - di Br PADAP in surfactant medium like pyridine in chloroform. Due to interaction of dye and surfactant in micellar medium it is observed that the colour of U(VI) - DBM - 3, 5, di Br PADAP system is sensitized further by increasing the colour sensitivity accompanied with red-shift.

U (VI)-DBM-3,5 di Br PADAP mixed ligand complex showed maximum absorbance at 590 nm and the reagent blank as no significant absorbance at 590 nm. U (VI)-DBM exhibits maximum absorbance at 400 nm when measured against its reagent blank. The U (VI) - DBM - 3,5 di Br PADAP system in 15% pyridine +10% n-butnol in chloroform medium is found to obey Beer's law upto Mg/ml of uranium (VI). The molar extinction coefficient ($\in$) and specific absorptivity (a) for (U) (VI)-DBM-3,5 di Br PADAP mixed ligand complex into 15% pyridine + 10% n-butanol in chloroform mixed solvent medium are found to be $\in = 9.1 \times 10^4$ l $mol^{-1}cm^{-1}$ and a =0.3824 mlg^{-1} cm^{-1}. The Sandell's sensitivity of the colour reaction is found to be s = 0.00262 $\mu g/cm^2$. It is also observed from Ayre's plot and Ringbom's Plot that the optimum range with minimum photometric error is in the region 1.6-4.0 μg/ml of uranium (VI) with a layer depth of one cm. The precision of the method is studied by analysing 50 ml sample solution of uranium (VI) varying from 4-100 ppm. The average standard deviation and coefficient of variation values are found to be 0.38 and 2.1 respectively.

The U(VI) - DBM - 3,5 di Br PADAP method is highly selective when the extraction is carried out in presence of masking mixture of EDTA and sodium fluoride and salting-out agent mixture calcium nitrate and aluminium nitrate. The co-extraction of many associated elements is prevented in presence of masking mixture in the initial extraction of uranium with DBM. However, anions like arsenate, carbonate phosphate, tartrate and citrate have low tolerance limits while the interference of thorium persists even at low costs concentrations because of its co-extraction.

While this method can tolerate upto 50mg chloride, higher concentrations only decrease the sensitivity slightly which can be overcome by using the standard Beer's law curve constructed in that particular chloride concentrations. Detailed investigations revealed that the present method can be applied for the precise estimation of uranium in various types of water samples of varying salinity.

The applicability of the present method has also been demonstrated for the analysis of diversified matrix materials by actually analysing uranium (VI) (100mg) in presence of likely associated metals Pb, Bi, Cu, As, Sn, Al, Fe, Ni, Co, Cr, W, Ti, Th, Be, Ca, Mg in the range 5-100mg.

Thorium interferes seriously in the present method when it is present in large quantities, because of co-extraction of Th(IV) with dibenzoyl methane even in presence of masking agents. However, detailed investigations reveal that interference of thorium can be removed by carrying out pre-extraction of uranium with ethyl acetate in presence of EDTA and aluminium nitrate and back extracting any thorium co-extracted by scrubbing the ethyleacetate extract with EDTA and aluminium nitrate. In the process of scrubbing it is observed that thorium is preferentially removed from the ethylacetate extract into the aqueous phase. The uranium in the organic phase can then be determined precisely after evaporating the extract ~~and~~ following the U(VI) - DBM - 3,5 -di Br

PADAP method. 5-200 μg of uranium in presence of 1g of thorium nıtrate are estimated by following the above procedure.

The applicability of U(VI) - DBM - 3,5 - di Br PADAP method for the analysis of low concentration of uranium in minerals, ores, rocks, soils and alloys has been demonstrated and procedures have been optimised for the dissolution of the above samples which re adoptable for the U(VI) - DBM - 3,5 - di Br PADAP extraction spectrophotometric method.

In chapter 4, the results obtained in the solvent extraction studies of uranium (VI) with DBM into pyridine + n-butanol in chloroform mixed solvent in the presence of masking and salting-out agents and spectrophotometric studies of U(VI) - DBM - 3,5 - di Br PADAP ternary system and its analytical applications are critically reviewed. Based on the observed data the mechanism of extraction of uranium-DBM into mixed solvent system and the facile interaction of uranium-DBM with 3,5 di Br PADAP in the surfactant medium are explained.

Section B includes chapters 5 to 9 which present the data on detailed investigations on the selective solvent extraction of thorium (IV) with TOPO and sensitive spectrophotometric estimation by Th(IV) TOPO-3,5 - di Br PADAP system.

In chapter 5 various extraction spectrophotometric methods on thorium reported earlier in literature and their application for analysis of different samples are briefly reviewed. Systematic and detailed investigations have been taken up on the extraction of TH(IV) with TOPO and on the spectrophotometric determination of thorium by the reaction of Th (IV) - TOPO with 3,5 - di Br PADAP in non-aqueous medium and these results are presented in chapter 6 and 7 respectively.

In the present investigation, it has been observed that thorium (IV) can be selectively extracted with TOPO in toluene at 3-7 M nitric acid medium as a colourless complex which is free from many interferences from many associated ions like Fe(III), Bi(III), Cu(II), Mo(VI), Mn(II), Ca(II), Mg(II), Pb(II), As(III), Co(II), etc.

Among several solvents tested as diluents for TOPO, toluene, carbontetrachloride and cyclohexane are found to be best solvents for the quantitative extraction of thorium, while toluene is found to be better facile medium for the ternary reaction of Th(IV) - TOPO system with 3,5- di Br PADAP. A 2000 fold of TOPO is found to be necessary for attaining quantitative extraction of Th and at low acidity 0.4 M, the extraction of Th is independent of TOPO concentration above the minimum level (2000 fold). Critical and detailed studies on the extraction of Th(IV) with TOPO under varied experimental conditions indicated that the extraction of Th(IV) with TOPO at low acidities can be represented as

$$Th(IV) + 4\,No_3 + 3\,TOPO \rightleftharpoons Th\,(No_3)_4 .\,3\,TOPO.$$

While at higher acidities an adduct like [Th (IV) $(No_3)_4$ 3 TOPO] n HNo_3 TOPO is responsible.

It is observed that the colourless Th(IV) - TOPO species extracted can be converted into an intensely coloured one by adding a chromogenic complexing reagent in non-aqeous medium. Of the various reagents PAR, PAN, TAR and 3,5 - di Br PADAP tested 3,5 - di Br PADAP gave high sensitivity colour reaction with good precision in non-aqeous medium.

Thus the extraction of thorium from high acidic medium with TOPO into toluene and addition of 3,5 di Br PADAP in ethanol + pyridine non-aqueous medium makes the reaction more selective and sensitive. It is observed that the presence of pyridine in the reaction medium accelerates the colour formation of Th (IV) - TOPO with 3,5- di Br PADAP and increases its sensitivity. It has been observed that a 30 fold excess of 3,5-di Br PADAP in ethanol is sufficient for attaining maximum absorbance.

Th(IV) - TOPO - 3,5 - di Br PADAP system obeys Beer's law upto 4.0μg and exhibits molar absorptivity of 7.7×10^4 l mol^{-1} cm^{-1} at 575 nm which is higher than those earlier reported with oxine ($\in$ = 10,000), molybdophosphoric acid ($\in$ = 33000), bromopyrogallol red ($\in$ = 51000), 4-(2- thiazolylazo) resorcinol ($\in$ = 31500) and xylenol orange ($\in$ = 55000). The Sandell's sensitivity of the colour reaction is found to be 0.00193 μg cm^{-2}. From the Ayre's and Ringbom's plots the optimum range with minimum photometric error for the estimation of thorium is found to be in the range 0.3μg to 2.0μg per ml of thorium.

The precision of the method is evaluated by the analysis of different concentrations of thorium in the range 30μg to 190μg which gave results with standard deviation 0.3 to 3.5 and coefficient of variation ranging from 2.0 to 2.4.

Job's method of continuous variations indicated ratio of 1:2 for Th: 3,5-di Br PADAP in the ternary species. Thus from solvent extraction data and Job's curves the final Th (IV) - TOPO - 3,5 - di Br PADAP ternary system is found to possess composition of Th: TOPO 3,5 di Br PADAP in the ratio 1:3:2.

The systematic and detailed studies on the analytical application of Th(IV) - TOPO - 3,5 - di Br PADAP method are given in chapter 8.

Th(IV) - TOPO - 3,5 -di Br PADAP can be directly applied for the precise analysis of Th(50-200 μg) without any pre-extraction of matrix, even in presence of moderately large concentrations of Fe(III) (100 mg), Mg(II) (15 mg), Ca(II) (2 mg), Al(III) (7 mg), Mn(II) (2 mg), Zn(II) (2 mg), Cu(II) (2 mg) and Pb(II) (2 mg).

However, for higher concentration of matrix components this method can be applied after bulk separation of the matrix components by pre-extraction. Thus, Th(IV) - TOPO - 3,5 -di Br PADAP method for the analysis of trace levels of Th in natural and industrial synthetic materials of diversified complex matrices can be applied after bulk separation of the matrix components by pre-extraction adopting one of the procedures given: (1) Bulk separation of matrix and constituent elements from thorium by pre-extraction with DDTC. This procedure is applied for the analysis of synthetic simulated samples of steels, copper base alloys, magnesium base alloys and nickel base alloys. (2) Selective pre-extraction of thorium with 4 methyl - 1,3 pentene - 2 - one (mesityl oxide) from 2 M HNO_3 medium and aluminium nitrate as salting-out agent. This procedure can be applied for the analysis of thorium in natural materials like monazite sand and aluminium alloys. (3) Bulk separation of titanium, iron, uranium and zirconium matrices by pre-extraction with aliquat 336. This method can be applied for the analysis of rutile, ilmenite samples and titanium base alloys. These procedures are applied either individually or in combination with another.

The capability of the above optimised procedures for the precise analysis of Th at trace level when present in various diversified complex matrices has been demonstrated by analysing spiked amounts of Th in various simulated synthetic sample solution of similar composition like cooper base, nickel base, magnesium base, aluminium base and titanium base alloys of actual BCS, NBS standards and natural materials like sand, rutile, ilmenite, etc.

In the results and discussion, chapter 9, the various experimental results obtained in the solvent extraction and spectrophotometric studies of Th(IV) TOPO - 3,5-di Br PADAP system are summerised and explained. The potentiality of the present method over other practised methods are compared critically. The various conclusions drawn based on the entire investigations are also summerised.

The results found in this research work will help many research scholars and scientists besides chemists in analysing further the properties of uranium and thorium and in identifying the relative efficiency of various analytical methods.

Dr. T.A.S. Devraj
Dr. D. Bhaskara Rao

1

Introduction

1.1 Introduction

In view of the rapidly expanding technological programmes, for the exploitation of nuclear energy under stringent environmental protection regulations, research studies on the development of precise analytical methods for the trace determination of uranium in diverse complex matrix materials became a demanding task for the analytical chemist.

The approximate average concentration of uranium in the rocks of the earth's crust is 4γ per gram. This is greater than that of many elements, including bismuth, cadmium, iodine, mercury and silver. The range of uranium concentration varies with the type of rock. Pegmatites (e.g. granite), the high-silica igneous rocks, contain of the order of 25γ per gram.

In general, the compositions of uranium-bearing samples are complex. In processable ores, uranium is found with such elements as niobium, vanadium, thorium, Zirconium and the rare earths, and in the nuclear fuels it is mixed with metals like zirconium, molybdenum, and various stainless steels. Analytical laboratories in plants recovering uranium from spent fuel are confronted with solutions of extremely high radiation levels containing fission products at sufficiently high concentrations to produce interference in quantitative methods for uranium. The recovery of uranium from expanded fuel for reuse results in a continually changing isotopic composition, so that only uranium separated from natural sources.has a predictable value.

The concentration of uranium in sea water is approximately 1γ per liter. This is an estimated reserve of 10^{10} tons of uranium, although, it is of doubtful economic usefulness. Many stream waters contain uranium at about this same level and constitute a geological means of tracing uranium deposits of leachable uranium. The presence of complexing anions, such as carbonate and chloride, in nature prevents hydrolytic precipitation of uranium.

Vegetation and some primitive forms of living matter contain uranium varying from 1×10^{-5} to 1γ per gram. The significance of uranium to cellular function is not known.

Alloys of uranium with Zirconium, aluminium, stainless steels, zirconium-niobium, aluminium-nickel, and molybdenum clad with alloys of these same elements, are presently used for metallic fuel elements. Uranium dioxide is an important nuclear fuel and is usually contained in tubes or cladding of zirconium stainless steel and aluminium. With increasing interest in reactors operating at high temperature, refractory fuels, consisting of mixtures of uranium oxides and carbides with ceramic materials are being considered. Fused fluoride salts and sulfuric acid solutions containing copper and nickel are used for homogeneous reactors. The analysis of expended fuel is further complicated by high concentration of fission and capture products. In the production of fissionable uranium-233 (by neutron capture on thorium -232), significant quantities of protoactinium and fission products are formed, producing another example of a difficult-to-analyze sample. Being the soluble from, the U (VI) [a potential and serious environmental pollutant] concentration levels in surface and ground waters are increasing with increasing mining activities and disposal of nuclear wastes etc., necessiating precise methods for monitoring uranium in various types of surface and ground water.

The chemistry of uranium is some what similar to that of chromium, molybdenum and tungsten, but it resembles more closely that of the actinides, neptunium and plutonium. Hence a separation usually precedes its determination.

Uranium often is separated by solvent extraction, then is determined by fluorescence, spectrophotometric, various types of electro analytical and gravimetric methods.

Uranium occurs in the +III, +IV, +V and +VI oxidation states, compounds of uranium (VI) and (IV) being of major importance. The yellow uranyl cation precipitates as the hydroxide $UO_2(OH)_2$ at PH ~ 4. Uranium (VI) is amphoteric, sparingly soluble uranate and diuranate (Na_2UO_4, $Na_2U_2O_7$) being formed in NaoH medium. Uranium (VI) gives peroxide, fluoride, tartrate, carbonate pH 7-12 and nitrate complexes. Uranium (IV) and thorium have similar properties. In acid medium sparingly soluble UF_4 and U $(C_2O_4)_2$ occur. Uranium (IV) is oxidized slowly to U(VI) by atmospheric oxygen and rapidly by iodine, iron (III) and stronger oxidizing agents.

1.2 Analytical Methods

Most analytical methods for uranium are adversely affected by foreign ions. This applies to samples containing fission products; the significant fission product nucleides range from elements with At. No. 32 to 65. Complex matrices also result from ore samples, with less than 1% uranium assay in the majority of samples. Alloys of uranium with special nuclear and high-temperature properties, containing corrosion - resistant elements such as zirconium, niobium and molybdenum are common place. Even though analytical measurements have improved in specificity, a separation step is usually required. Excellent analytical separation procedures are available for uranium, many of which are based on production processes.

The majority of the methods for the separation of uranium from interfering elements are based on the separation of uranium itself from solution being analyzed, and only a minority on the separation of the accompanying elements.

The methods of precipitation, extraction and chromatography are the most widely used. Distillation, sublimation, electrolysis, electrodialysis etc., are also sometimes employed. For the separation of small amounts of uranium, the most suitable methods are coprecipitation with a carrier and extraction, as well as chromatography.

The choice of method depends upon the method of determination and the composition of the solution being analyzed and mainly upon the nature of the interfering elements, and their concentration, the uranium content of the solution and the accuracy required. In many cases, the possibility of employing a given method has been considerably increased by the widespread use of different masking complexing agents in the separation of uranium from interfering elements.

1.3 Separation Techniques

Liquid - Liquid extraction is the most used separation techniques for uranium both for production processes and for analytical samples. For the small batch samples encountered in analytical work, ion-exchange is sometimes preferred if the equivalent of more than one contact is required to effect sufficient separation. A disadvantage of ion-exchange as well as of column or paper chromatography, is that the complete recovery of uranium required for high precision methods is difficult to achieve.

Factors contributing to the popularity of Solvent Extraction technique are high recoveries of uranium with one-batch contact, simplicity of apparatus, and specificity. Other advantages, compared to such techniques as ion-exchange and precipitation, are freedom from coprecipitation and adsorption problems, ready adaptation to remote operations, and better reproducibility between laboratories. Highly specific separations have been obtained with little additional manipulative effort by use of 2-cycle extraction system with the second of a different type than the first.

Organic acids, ethers, ketones, esters, alcohols, organic derivatives of phosphoric acid, and nitro paraffins have been used for uranium separation. The feature common to these is non-chelate oxygen bonding. This type of bonding is more specific than is chelate formation. In choosing an extraction system, criteria to be considered are specificity, high distribution coefficients, and solvent stability.

Of the oxygen bonding system, extraction of tetra-propylammonium uranyl trinitrate into methyl isobutyl ketone from an acid deficient aluminium nitrate-salted aqueous phase meets these criteria(1).

Tris (2-ethyl n-hexyl) phosphine oxide, a more stable but less specific extractant, is also generally applicable, especially when combined with EDTA masking.

1.4 Solvent Extraction

Uranium (VI) can be selectively extracted from nitrate medium with polar solvents. The extracted species is solvated undissociated uranyl nitrate. The concentration of free nitric acid must not be too high

(0.1-1M) because at higher acidities the degree of extraction of other metals (Ce(IV), Th, Zr, Au(III), As(V), Bi) is increased. To prevent the dissociation of uranyl nitrate and to enhance the distribution coefficient of uranium, considerable quantities of aluminium, calcium or ammonium nitrate (salting-out agents) are added to the aqueous solution. Sulphate, and small amounts of phosphate and fluoride do not interfere. In the presence of chloride, iron(III) and other metals giving chloride complexes can be extracted.

The solvents used include methyl ethyl ketone in Ccl_4[2], mesityl oxide,[3], ethyl acetate[4] TBP[5] TBP in isooctane[6], TBP in Xylene[7] and TOPO in cyclohexane[8]. Uranium (VI) has also been extracted from nitrate medium with triphenyl arsine oxide in chloroform[9], and as the tetrapropyl ammonium uranyl nitrate complex in MIBK[1].

In the presence of trioctylamine and other amines, uranium can be extracted with non-polar solvents from hydrochloric[10-13], sulphuric,[10,14-17] or acetic acid[18] media. Nemodruk[19] extracted the anilinium salt. Ichinose[20] has extracted uranium (VI) with MIBK from chloride medium (HCL + LiCl).

Other uranium complexes for its extraction include those with cupferron,[21] acetylacetone,[22] DDTC[23] , i-nitroso-2-naphthol[24], di-n-buyl arsinic acid[25] HDEHP[26], Phenylacetic acid[27] and N-Phenyl benzo hydroxamic acid[28].

1.5 Spectrophotometric methods for the Trace Analysis of Uranium.

Spectrophotometric methods are widely applied for the trace determination of uranium as the necessary equipment is simple and inexpensive . The basic advantages of spectrophotometry like high sensitivity, in some cases high selectivity, ease in application, simplicity of instrumentation, ready availability of reagents and instruments are well known. Spectrophotometric methods can be used for precise determination of practically all metals and non-metals in any material.

Among all instrumental analytical techniques spectrophotometric methods are the most chemical ones. They are particularly important as control methods in the preparation of certified standards where high precision and reliability of the results outweigh the disadvantage of the rather time-consuming analytical procedure.

The possibility to develop direct and rapid spectrophotometric analysis techniques depends mainly on two factors, selectivity and reliability of the reaction. Reliability in this case, is meant to be the indifference of the results to minor variations in the reaction conditions like pH, temperature etc., as well as the insignificant influence on the basic reaction when other substances present in the solution such as weak oxidisors, reducers and organic solvents, salts etc.

In the trace analysis of complex materials the matrix is some times a single element as in the analysis of high purity metals. The preconcentration of trace elements, can be carried out by solvent extraction or on ion exchanger columns or by coprecipitation with a collector (scavenger or by volatilization).

Preconcentration by solvent extraction plays double role in trace analysis. It enhances the relative concentration of the element to be determined, thereby increasing the sensitivity of the method. It separates the determined trace elements from macro components (matrix) which frequently interfere with the determination.

1.6 Precision, Accuracy and Selectivity

The precision of spectrophotometric methods depends on the concentration of the determined and varies from approximately 0.5 to 2% under suitable measuring conditions. The precision attainable is a function of the absorbance measured. The measuring errors are smaller if the measured absorbance is in the range 0.2 to 1.0.

Selectivity of the spectrophotometric methods usually depends on the nature of the reagent used, the oxidation state of the element, the pH of the medium, and the masking agents. If, in spite of use of all these factors, some elements can still interfere in a determination, the species to be determined has to be separated from the interfering elements or vice-versa.

The optimum pH range for reaction with a given element is related to the nature of the element and the reagent. For the reagents of the ROH type there is a relation between the colour reactions and the hydrolysis of given elements. In strongly acid solution the colour which readily hydrolize, i.e., Zr, Hf, Tl, Th, U(IV). In acid medium the metal ions of the hydrogen sulfide group that form the most stable sulfides react with RSH type of reagents.

Increased selectivity of spectrophometric methods is mainly obtained by masking the interfering ions. The most often used masking agents are EDTA, cyanide, citrate, tartrate, fluoride, iodide and thiosulfate.

High selectivity is obtained by a suitable combination of masking agents, pH and appropriate reducing or oxidizing agents. It should be noted that solvent extraction itself also result in an increase of selectivity. In general, the extractive spectrophometric methods are more selective than methods carried out in aqueous media. Combination of a selective extraction method with a sensitive spectrophotometric method of determination of poor selectivity is an example of advantageous analytical hybrid solution.

1.7 Colour System Used in Spectrophotometry

Sensitive spectrophotometric methods are based on colour systems formed from reactions of the elements to be determined. The majority are complexation reactions and rarely other types. Thus, organic complexing reagents play the most important role in trace spectrophotometric analysis[29-31]. The presence of some chromophors, such as azo or p-quinoid groups, in organic reagents, results in high colour intensity of the chelate compounds formed with these reagents. In addition to chelating reagents there exist also colour-forming, organic non-chelating reagents, i.e., the basic and acid dyes. These dyes form ion associates with the determined elements, i.e., anionic and cationic complexes. The more sensitive spectrophotometric methods are more and more often based on ternary systems involving basic dyes or some surfactants.

In the presence of some long-chain quaternary bases (showing surface active properties) and chelating triphenylmethane dyes, ternary complexes are formed with metal ions. These complexes show more intense colour than the respective simple complexes with the above mentioned chromophoric reagents[32-35]. In addition to a high increase in the colour intensity a bathochromic shift of the absorption maximum takes place, as a result of the different λ max of the complex.

The colour reactions of uranium can be divided into two classes. The first class comprises those reactions which are based on the chromophoric effect of uranium, i.e., on the ability of U^{4+} and UO_2^{2+} ions to form colored compounds with simplest inorganic or organic reagents. It is usually assumed that the chromophoric effect is actually produced not by the elements or the ions but rather by the atomic combinations U-O, U-S or possibly U-N[36-37]. Many other elements besides uranium give chromophoric reactions : Fe, Cu, Ni, Mn, Cr, V, Mo, Co, Ce, etc. They are all capable of forming coloured products with colorless reagents. On the other hand, Th, Zr, Hf, Al, Zn, Ca, Ba and others do not give chromophoric reactions and do not form coloured products with colorless reagents. Hence, these latter elements, at least, do not interfere with the colour reactions of uranium, which ensure a greater selectivity for this class of reactions

Examples of the chromophoric effect are supplied by the colour reactions or uranium with salicylic acid, KSCN, and resorcinol, as well as by the colour produced in concentrated solutions of sulfuric, perchloric and phosphoric acids.

The second class of reactions are the colour reactions of uranium with intensely coloured organic reagents. In this case, no new colour is produced, but the intensity of the colour already present increases. The end products of this type of coloured reactions are chelates of uranium with the molecule of the reagent. A change in colour is produced by any element which forms chelates with the reagent, whether it gives chromophoric reactions or not.

The most widely used reagents are azo dyes of type R_1-N=N-R_2, and the most suitable for uranium are the azo dyes containing the radical of arsenic acid :

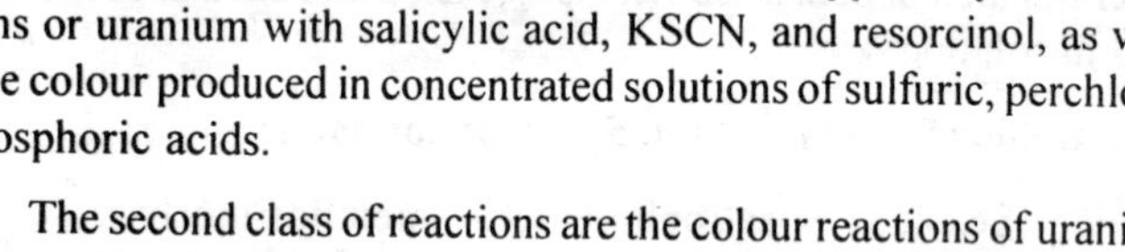

AsO_3H_2 OH OH N=N and AsO_3H_2 OH N = N

They form complexes in which the ions of U(IV) and U(VI) are very firmly bound, and the complexing reactions ensure high contrasts in colour[38-41]. Reagents of this group are not specific for uranium.

Although colour reactions with coloured reagents sometimes lack in selectivity, the sensitivity of some of them exceeds by 1 or 2 orders of magnitude that of the reactions based on the chromophoric effect of uranium. This is the principal feature of reactions of the second type.

It is not always possible to draw a distinction between the two types of reactions as the mechanism may sometimes be partly according to one type, and partly to the other, like in the colour reactions with alizarin S, Aluminon, nitroso-R-salt, morin, H_2O_2 and some others.

It is possible to determine uranium in more complicated substances by means of organic solvents, since the uranium can be separated from interfering elements by extraction. It is also possible to concentrate the uranium during the extraction. Some of the methods for the determination of uranium in organic solvents are connected with the analysis of industrial extracts.

1.8 Determination of Uranium from the Colour of its Compounds with Colourless and Slightly Coloured Organic Reagents

The reaction between uranium and colourless or slightly coloured organic reagents is based on the formation of chromophoric combinations: uranium-oxygen, uranium-sulfur, etc. As a rule, the sensitivity of the methods with these reagents is of the same order, about 5 μg/ml of uranium: their selectivity is relatively good, but a preliminary separation of uranium from the accompanying elements is always necessary.

The selectivity of the methods described is considerably improved by the use of complexones to bind the interfering elements. According to Pribil[42] Palei and Udal'tsova, and other authors[43-48] with complexones the determination of uranium by dibenzolymethane, diethyldithiocarbamate, morin or salicylaldoxime can be carried out in the presence of elements which would other wise interfere.

The following reagents have been used for the quantitative determination of uranium : ascorbic acid[49-51] salicylaldoxime[52, 53] salicylic and sulfo salicylic acids[49,52, 54-58] chromotropic acid[49, 59-63] R-salt and nitroso-R-salt[64] ammonium thioglycolate[51, 65-67], 8-hydroxyquinoline[49, 68-72] and its derivatives[72] dibenzoylmethane[73-81] thenoyltrifluoroacetone[82] resorcinol[83] tannin[84] quercetin[85] morellin,[86] and other [87-92].

Though a number of inorganic and organic reagents have been reported, only a few reagents from coloured complexes with molar absorptivity (e) greater than 2 x 10^4 l $mole^{-1}$ Cm^{-1} which is required for the estimation of uranium in the range of 1-100 ppm. Chromophoric chelating agents which yield greater sensitive colour reaction are :

4-((2-Pyridylazo) resorcinol) $\in$ = 3.87 x 10^4 l $mole^{-1}$ cm^{-1}
4-(2-thiazolylazo) resorcinol $\in$ = 3.3 x 10^4 l $mole^{-1}$ cm^{-1}

Arsenzo III $\in$ = 5.3 x 10^4 l $mole^{-1}$ cm^{-1}
Eriochrome Black T $\in$ = 3.0 x 10^4 l $mole^{-1}$ cm^{-1}

However, these reagents are not selective and require rigorous and elaborate experimental procedures for the actual application for the trace analysis of uranium in real diversified complex matrix materials. Similarly, a number of extracting agents like Pyrrolidine dithiocarbamate, TOPO, Hydroxamic acids, thiocyanate, dibenzoyl methane etc. are available. They do not produce sensitive colour reactions with uranium and in many cases these reagents are also not selective to uranium and subject to interference from various associated elements.

The author in his pursuit to develop highly selective extraction and sensitive spectrophotometric methods for the precise estimation of uranium in various natural intermediate and finished products and effluents etc., studied the conditions for selective extraction of uranium using dibenzolymethane and its sensitive estimation by interaction of the extracted chelate system with a chromophoric chelating agent like 2-(3,5-dibromo-2-Pyridylazo)-5-diethylamino Phenol (3,5-diBr PADAP) in a non-aqueous medium.

Detailed and systematic studies were made for developing highly selective and simple procedures for the extraction and preconcentration of uranium from various complex matrix components by studying the extraction process in presence of suitable combination of masking agents with proper pH control.

Similarly, the author also made detailed studies on the effect of surfactants like pyridine, quaternary ammonium compounds on the uranium - dibenzoyl methane - 3, 5 diBr PADAP complexes to achieve higher colour sensitivity, as metal complexes formed in micellar medium are more stable and produce red shifts in absorption bands with increase in molar absorptivity. To make the discussion of the results more lucid, the physico chemical characteristics of the dibenzoylmethane and 3, 5 diBR PADAP are discussed in the following pages.

1.9 Dibenzoylmethane

Dibenzoylmethane belongs to the class of

β-diketones R-CO-CH_2-CO-R

which in their enol form R-C(OH)=CH-CO-R are capable of complexing with certain cations. The sensitivity of these reagents increases in the following sequence :

Acetylacetone (CH_3-CO-Ch_2-CO-CH_3)
Benzoylacetone (C_6H_5-CO-CH_2-CO-CH_3)
Dibenzoylmethane (C_6H_5-CO-CH_2-CO-C_6H_5)

The coefficient of molar extinction is 580 for acetylacetone,[93] and 17, 810 for dibenzoylmethane. The solubility of these reagents in water decreases when an aromatic radical is substituted for the alkyl, and hence dibenzoylmethane is totally insoluble in water, in contrast to its analogs.

Dibenzoylmethane has been studied as a reagent for the determination of sub-milligram amounts of uranium since 1953. Yoe, Will and Black[81] studied its general properties and used ether extraction to spare uranium from interfering elements.

Yamane[74] suggested a method for the determination of uranium in alcoholic solutions with certain B-diketones, including benzoylacetone, dibenzoymethane, isonicotyl dibenzoylmethane, 2-furoyl dibenzoylmethane and o–, m–, and p-methoxydibenzolymethane. Isidate and Yamane[73] also studied other substituted β-diketones : o–, m–, and p-nitroso-dibenzoylmethane, m– and o–aminobenzoylmethane : the compound of the latter with uranium has a high coefficient of molar extinction ε_4 10 mμ = 30,000 in alcoholic solutions.

1.10 2-(2-Pyridylazo)-5-Diethylaminophenol (PADAP)and its Halo Derivatives

Pyridylazo compounds with a hydroxy group in the ortho position with respect to the azo group are valuable chromogenic reagents. They have been widely used in complexometric and extraction photometric methods for the determination of many ions. Highly sensitive reagents have been obtained by the introduction of the disthylamino group, which is a strongly electron-donating group in the para position next to the azo group.

2-(2-Pyridylazo)-5-diethylaminophenol (PADAP) is a dark-red solid melting at 109°C and soluble in ethanol, chloroform, acetone, cyclohexane, and ether. 2-(5-bromo-2-Pyridylazo)-5-diethylaminophenol (5-Br-PADAP) is a brick red solid melting at 151°C and soluble in ethanol, acetone, chloroform, ether and dioxane.

The azo dye, 2-(2-Pyridylazo)-5-diethylaminophenol (PADAP) and its bromine substituted derivative, 2-(5-bromo-2-Pyridylazo)-5-diethylaminophenol (Br-PADAP), the latter having a greater tolerance to iron(III), have found wide application as very sensitive reagents in the spectrophotometric determination of uranium(VI).

2-(3,5-dibromo-2-Pyridylazo)-5-diethylaminophenol (3,5-diBr PADAP) is a dark red solid melting at 155°C and soluble in ethanol,

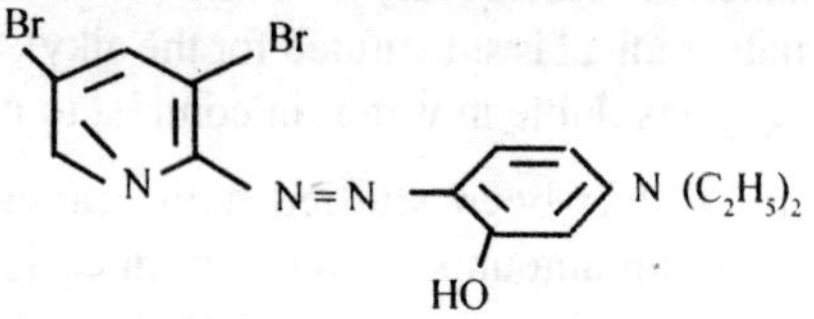

Br Br

N N=N N $(C_2H_5)_2$

HO

chloroform, acetone, benzene and ethyleane dichloride.

These reagents react with many metal ions and can be used as metal indicators in complexometric titrations and as photometric reagents[94]. Known colour reactions of metal ions with these compounds in aqueous media are summarised in Table 1.1.

TABLE - 1.1
Colour Reactions of Metal Ions with PADAP, 5-BROMO PADAP and 3, 5 Dibromo PADAP.

Metal Ion	*P A D A P*		*5-Bromo-PADAP*		*3,5-dibromo-PADAP*	
	Colour	*pH*	*Colour*	*pH*	*Colour*	*pH*
1	2	3	4	5	6	7
Ba (II)	Rose	9.5-10	Rose	7.5-8	Darkred	6.5-7
Bi (III)	Violet	1.5-2	Violet	1-1.5	Violet	0.65-1
Ca (II)	Rose	9.5-10	Darkred	8.5-9	Darkred	8.5-9
Cd (II)	Rose	5.5-6	Rose	4.5-5	Rose	4-4.5
Co (II)	Darkred	2.5-3	Darkred	2-2.5	Darkred	1.5-2
Cr (III)	Darkred	2.5-3	Violet	2-2.5	Violet	2-2.5
Cu (Ii)	Darkred	2-2.5	Darkred	1-1.5	Darkred	0.65-1
Fe (II)	Darkred	2-2.5	Darkred	2-2.5	Darkred	1.5-2
Fe (III)	–	–	Darkred	2-2.5	Darkred	2-2.5
Ga (III)	Darkred	2.5-3	Darkred	1-1.5	Darkred	1.5-2

Contd.

TABLE - 1.1 (Contd.)

1	2	3	4	5	6	7
Hg (I)	Rose	1.5-2	Darkred	1.5-2	Darkred	0.65-1
Hg (II)	Rose	2-2.5	Darkred	2-2.5	Darkred	1.5-2
In (III)	Darkred	2-2.5	Darkred	1.5-2	Darkred	1.5-2
Mg (II)	Rose	9.5-10	Darkred	8.5-9	Darkred	8.5-9
Mn (II)	Rose	5.5-6	Darkred	3.5-4	Darkred	3.5-4
Nb (V)	Violet	2.5-3	Blue	0.65-1	Blue	0.65-1
Ni (II)	Darkred	4-4.5	Darkred	2.5-3	Darkred	1.5-2
Pb (II)	Darkred	4.5-5	Darkred	3-3.5	Darkered	2-2.5
Pd (II)	Violet	0.65-1	Violet	0.65-1	Violet	0.65-1
Sb (III)	Darkred	4-4.5	Darkred	0.65-1	Darkred	0.65-1
Sc (III)	Red	5-5.5	Darkred	4.5-5	Darkred	4.5-5
Sr (II)	Rose	4-4.5	Rose	4.5-5	Rose	4-4.5
Sr (II)	Rose	9.5-10	Rose	8.5-9	Rose	9.5-10
Th (IV)	Darkred	4.5-5	Rose	3-3.5	Rose	4-4.5
Ti (IV)	Darkred	4-4.5	Violet	3.5-4	Violet	4-4.5
Ti (III)	Darkred	1.5-2	Darkred	2-2.5	Violet	0.65-1
UO_2^{2+}	Darkred	2.5-3	Darkred	2-2.5	Violet	1.5-2
V (V)	Violet	1.5-2	Blue	1-1.5	Green	0.65-1
Zn (II)	Rose	4.5-5	Rose	4-4.5	Violet	2.5-3
Zn (IV)	Darkred	4-4.5	Violet	1-1.5	Violet	0.65-1

In the use of 3,5 diBr PADAP as a photometric reagent, the working conditions like pH, excess of 3,5 diBr PADAP, metal concentration range, ionic strength and buffer species must be selected carefully.

1.11 Extractive Spectrophotometric Methods

Extractive spectrophotometric methods play a very important role in trace metal analysis of complex materials as they can be made very selective and sensitive by properly modifying and optimising experimental conditions such that they are suitable for the analysis of a particular analyte in a specific matrix. By solvent extraction we can achieve preconcentration and selective separation of analyte from the matrix components. Further, by using a suitable combination of masking agents the selectivity, and using a chromogenic chelating dye 3, 5 di Br PADAP in a surfactant medium, the sensitivity, can be adopted for the precise analyte analysis at much lower concentrations and in presence of complex matrices.

The 2-pyridylazo compounds are selective neither as chromogenic agents nor as extractants though they give intensely coloured reactions. However, some measures can be taken to improve the situation. The most important ones are adjustment of the pH and the use of masking agents. Hydrogen ion competes with the metal ion for combination with the 2-pyridylazo compound. Consequently, the higher the stability of the metal complex, the lower the pH at which it can exist. Therefore, at lower pH only a few number of metals will be complexed. The pH control in combination with masking agents will be useful both for selective colour formation and extraction.

The use of 2-pyridylazo compounds as chromogenic agents can be frequently combined with solvent extraction techniques by the following methods :

1. The azo regent is added to the aqueous solution, and the chelates formed is extracted into an organic phase and is then subjected to spectrophotometric measurements.
2. The metal is first extracted from the aqueous solution into an organic phase by means of a suitable organic extractant, and the azo reagent is then added to the organic phase. The colour developed is evaluated photometrically.

The increased selectivity of metal - 3,5 diBr PADAP spectrophotometric methods can be mainly obtained by masking the interfering ions by transforming them into stable complexes so that they can neither be extracted along with the analyte nor react with 3,5 di Br PADAP. High selectivity is obtainable by suitable combination of masking agents and pH, as the effective stability of complexes varies with pH and presence of other complex forming substances. Thus, even when the reagent is not selective the method can be specific for the determination of analytes, by using suitable combination of masking agents and pH. Wide variety of complexing agents, such as EDTA, cyanide, tartrate, citrate, thiosulphate etc., which form highly stable colourless complexes with a number of metal ions can be used as masking agents.

The addition of a surfactant to a metal dye complex may lead to red shifts in absorption bands and increases its molar absorptivity. Usually the metalchelate complexes formed in micellar system are more stable than those formed in the absence of micelles. Miscelles are responsible for many practical applications in spectrophotometric analysis owing to their effect on the coloured complexes like (i) altering the micro-environment, (ii) catalysis by concentration effect in the micellar pseudo phase, and (iii) alteration of reaction path-ways, rates and equilibria with optically transparent and stable systems.

In the present investigation the author has studied the application of U(VI)-dibenzoylmethane-3,5 di Br PADAP colour reactions for the precise analysis of U(VI) at trace level in widely varied complex matrices by selectively extracting in presence of a combination of masking agent and sensidizing these colour reactions using dye-surfactant interaction in a micellar medium.

References

1. Maeck, W.J., G.L. Booman, M.C. Elliott, and J.E. Rein *Anal. Chem.*; 30, 1902; 1958.
2. Kuznetsov, V.I. and Blekhta, V. Collection Czech, *Chem. Commun.*; 26, 1092; 1961.
3. Dhara S.C., and Khopkar, S.M., *MiKrochim. Acta*; 931; 1965.
4. Guest, R.J and Zimmerman J.B.; *Anal. Chem.* 27, 931; 1955.
5. Eberle, A.R and Lerner, M.W., *Anal. Chem* 29, 1134; 1957.
6. Francois, C.A., *Anal. Chem.*, 30; 50; 1958.
7. Nemodruk, A.A and Glukhova, L.P.; Zh. *Analit Khim.* 21, 688; 1966.
8. Horton, C.A and White, J.C. *Anal. Chem.* 30, 1779; 1958.
9. Keil R.Z, *Anal. Chem.* 244, 165; 1969.
10. Tikhomirov, V.I., Kuznetsova A.A and Batorovskaya, E.D. *Radio Khimiya* 6, 173, 182, 187; 1964.
11. Hodara, I and Balouka, I. *Anal. Chem.* 43, 1213; 1971.
12. Vieux, A.S., *Bull Soc. Chim.* 4281; France; 1968.
13. Onishi, H. and Sekine, K. Talanta 19, 473; 1972.
14. Boirie, C. Bull *Society Chim*, 1088, France; 1958.
15. Deptula, C. and Minc, S.J. *Inorg. Nucl.* Chem., 29,221; 1967.

16. Gagliardi, E. and Ilmaier B. *Mikro Chim. Acta*, 1259; 1968.
17. Juznic, K. and Fedina S., *Mikro Chim. Acta*, 39; 1974.
18. Moore, F.L. *Anal. Chem.* 30, 908; 1958 and 32, 1075; 1960.
19. Nemodruk, A.A. Tr. Komis Po *Analit Khim*, Akad Nauk SSSR, 14, 141; 1963.
20. Ichinose, N., *Talanta* 18, 21; 1971.
21. Rulfs, C.L.De, A.K. and Elving, P.J. *Anal. Chem.*, 28, 1139; 1956.
22. Krishen, A. and Freiser, H., *Anal. Chem*, 29, 288; 1957.
23. Clayton, R.F. Hardwick, W.H. Moreton Smith, M. and Todd, R. *Analyst* 83, 13; 1958.
24. Alimarin, I.P. and Zolotov, Yu, A., Zh. *Analit. Khim* 12, 176; 1957.
25. Pietsch R and Pichler, R., Z. Anal. Chem. 190, 319; 1962.
26. Schmid, E.R. and Pfannhauser, W., *Mikro Chim. Acta* 434; 1971.
27. Adam, A. and Pribil, R. Talanta 20, 1344; 1973.
28. Shukla J.P. Agrawal, Y.K. and Bhatt, K., *Sepn. Sci.* 8, 387; 1973.
29. Sandell, E.B. and Onishi, H., *Photometric Determination of Traces of Metals, General Aspects*, 4th Ed., John Wiley and Sons, New York, 1978.
30. Savvin, S.B. Crit, Rev. Anal. Chem., 8, 55; 1979.
31. Holzbecher, Z. Divis, L. Kral, M. Sucha, L. and Vlacil, F., *Handbook of Organic Reagents in Inorganic Analysis*, Horwood Wiley, Chichester, England, 1976.
32. Svoboda, V. and Chromy, V., *Talanta* 12, 431, 437, 1965 and 13, 237; 1966.
33. Dagnall, R.M., West. T.S. and young, P. Analyst; 92, 27; 1967.
34. Bailey, B.W., Chester, J.E., Dagnall, R.M. and West, T.S., Talanta, 15, 1359; 1968.
35. Somner, L. and Kuban, V., *Collection Czech. Chem. Commun*; 32, 4355; 1967.
36. Kuznetsov V.I *Uspekhi Khimii*, 21, 175; 1952.
37. Muller A. - Z *Prakt. Chem.* 157, 89; 1940.
38. Kuznetsov V.I. Doklady *A.N. SSSR*, 31, 895; 1941.
39. Kuznetsov V.I. Doklady *A.N. SSSR*, 52, 39; 1946.
40. Kuznetsov V.I *ZhAKh*, 13, 220; 1959.
41. Kuznetsov V.I. and S.V. Savvin - *Radiokhimiya* 1, 589, 1959.
42. Pribil, R. and J. Vorlicek - *Chem. Listy.* 46, 216; 1952.
43. Carter, J.A and C.W. Weber - *U.S. Atomic Energy Commn. Rep. TTD.* 7516, 186; 1956 cited from *Anal. Abstr.* 5, 490, 1958.
44. Majumdar A.K. and J.B. Chowdhury - *Anal. Chim.* Acta, 19, 576; 1958.
45. Sen SARMA R.N. and A.K. Mallik - *Anal. Chem.* Acta, 12, 329; 1955.

46. Sen SARMA R.N. and A.K. Mallik - *Science and Culture*, 20, 135; 1954.

47. Sen Sarma R.N. and A.K. Mallik - Z *Anal. Chem.*, 148, 179; 1955.

48. Shelton, R.A.J. - *Analyst* 82, 531; 1957.

49. *Anal. Chem. of Uranium and Thorium* Russian Translation (P.N. Palei, Ed.) 1956.

50. Palei, P.N. - *Peaceful Uses of Atomic Energy, Proceedings of the International Conference in Geneva*, 8; 1955.

51. Dizdar, Z.I. and I.D. Obrenovic - *Second United Nations International Conference of the Peaceful uses of Atomic Energy* R/471; 1958.

52. Kuznetsov, V.I. - *Research on Color Reactions for Inorganic Ions*, Doctorate Thesis; Moscow, Geokhi AN SSR, 1948.

53. Almassay, G. - Magyar Kem, Folyoirat, 61, 404; 1955, cited from *Anal Abstr* 3. 1705; 1956.

54. Babko, A.K. - *Zapiski Institution Khimii* AN UKr SSR, 8, No. 2, 1946.

55. Tret'yak, Z.A., *Dissertatsiya* (Thesis) - Khar'Kovskii, Gosudarstvennyi Universitet im.A.M. Gorkogo, 1953.

56. Beck M.T. and E. Hantos - *Acta. Chem. Acad. Scient.* hung, 8, 233; 1955.

57. Foley R.T and R.C. Anderson, *J. Am. Chem. Soc.* 71, 909; 1949.

58. Thomason, P.E. and L.A. Nutting, U. Koskela and W.M. Byerly *U.S. Atomic Energy Commn. Rep.* ORNL - 1641, 15, 1955, cited from Anal. Abstr. 4, 2183; 1957.

59. Bhaduri A.S and P. Ray - Z *Anal. Chem.*, 154, 103; 1957.

60. Mukherji A.K. and A.K. Dey, *J. Indian Chem.* Soc., 35, 113; 1958.

61. Sarma, B. and C.P. Savariar - *J. Sci. Ind. Research*, 16, No. 2, 80, 1957.

62. Smith, H.R.D. Z - *Anorg. Chem.* 150, 163, 1936.

63. Tanajev N.A. and G.A. Panschenko - *Z Anorg. Und Allgem. Chem.* 150, 163. 1926.

64. Rao M.N. and RBS Rao - *Z Anal. Chem.*, 142, 161; 1954.

65. Dizdar, Z.I and I.D. Obrenovic *Analyst*, 83, 177, 1958.

66. Swank, H.W. and M.G. Mellon - *Ind. and Eng. Chem. Anal.* Ed., 10, 7, 1938.

67. Woodman, F.J., T.G. Clinton, W. Fletcher, and G.A. Welch - *Second United Nations International Conference on the Peaceful Uses of Atomic Energy*, A/Conf. 15/P/1453, 1958.

68. Tinovskaya, E.S. - *Zavodskaya Laboratoriya*, 17, 387, 1951.

69. Bullwinkel, E.P. and P. Noble - *J. Am. Chem. Soc.*, 80, 2955, 1958.

70. Hok, B. Svensk. Kem, Tidskr. 65, 106, 1953.

71. Kibby, K. and R. Granlly, *Anal. Chim. Acta*, 19, 363, 1958.

72. Rulfs, C.L. A.K. De, J. Lakritz, and P.J. Elving, *Anal. Chem*, 27, 1802, 1955.

73. Isidate and Yamane - J. Pharmac. Soc. Japan, 77, 386, 391, 396; 1957.

74. Yamane *J. Pharm. Soc.*, Japan, 77 400; 1957.

75. Adams, J.A.S. and W.J. Maeck - *Anal. Chem.* 26, 1635, 1954.

76. Blanquet, *P. Anal. Chim. Acta.*, 16, 44; 1957.

77. Francois, C.A. - *US Atomic Energy Comm. Rep.* Dow-150; 1956 cited from Anal. Abstr. 4, 3939; 1957.

78. Fancois, C.A. - *Anal. Chem.* 30, 50; 1958.

79. Horton, C.A. and J.C. White - *Anal. Chem.* 30, 1779; 1958.

80. Pribil, R. and M. Jelinek, *Chem. Listy.* 47, 1326; 1953.

81. Yoe, J.H.F. Will and R.A. Black - *Anal. Chem.* 25, 1200; 1953.

82. Khopkar, S. - *A.de Chem, and Ind.* P. 291; 1959.

83. Jain, P.C. and G.S. Rao - *Anal. Chim.* Acta, 20, 171; 1959.

84. Das Gupta P.N. - *J. Indian Chem. Soc.* 6, 763; 1929.

85. Komenda - *J Chem. - Listy*, 47, 531; 1953.

86. Rao, B.R. and C.C. Patel - *Proc. Indian Acad.* of Sci., 43, 276; 1956.

87. Sandell E.B. Colorimetric *Determination of Traces of Metals*, Interscience, 1944.

88. Bandopadhayay, D. - *J Indian Chem. Sc.*, 33, 269; 1956.

89. Chakrabartty, A.K., D. Sen. and P. Ray - *J Indian Chem. Soc*, 30, 491; 1953.

90. Kantharajurs, M. and K. Neela Kantam - *J Sci. and Ind. Res.* 11, 78; 1952.

91. Srinivasulu, K.D. Purushottam and B.R. Rao - Z *Anal. Chem.* 159, 406; 1958.

92. Szarvas, P. Z. Jarabin and L. Dede Magyar *Kem. Folyoirat*, 63, 151; 1957.

93. Jerowska - Trzebiatowska and A. Bartecki - Bull de Acad. Polon. *des Sciences. Chem. Ser.*, 6, 41; 1958.

94. S.I. Gusev and L.M. Shchurova, Zh. Analit. Khim., 21, 1042; 1966.

2

Studies on the Solvent Extraction of Uranium(VI) with Dibenzoylmethane

2.1 Introduction

Determination of uranium in the presence of different mineral impurities is a difficult analytical problem, especially as the uranium content being determined is much smaller than the content of impurities coming from the different minerals present and the solutions used for their decomposition. Methods for determination of uranium in phosphate rocks and aqueous solutions obtained by their decomposition with mineral acids are also known. The methods are often based on solvent extraction with reagents such as tributyl phosphate or trialkylamines.

Many of the analytical methods for uranium, making use of various laboratory and instrumental techniques have already been published[1-10], but so far none of them is universal enough to be directly applied to determination of uranium in specific materials, like sulfuric acid, aluminum nitrate solutions, protective materials used in nuclear power plants, port-land cement, sea bottoms, rocks and minerals, uranium ores, soils, sea water and natural water.

From the work of Hardwick and Smith[8] an analytical separation procedure is available for extraction of traces of uranium in irradiated thorium into isobutyl-methyl Ketone (IBMK) from a dilute nitric acid solution saturated with ammonia nitrate. However, the separation seems to be incomplete, because the thorium is slightly extracted. According to Peppard[11] et.al, Ishimori[12] et.al and Ishii and Takeuchi[13] in the extraction of thorium and uranium with tributyl phosphate from hydrochloric acid solution, thorium is partly extracted together with uranium and the complete separation of these two elements is impossible.

Norio Ichinose[14] reported that the extraction of uranium from hydrochloric acid medium can be markedly enhanced by using lithium chloride or magnesium chloride as salting out agents.

Thus, though the extraction of uranium, thorium and other associated elements by isobutyl methyl ketone (IBMK) from nitric acid medium in presence of various nitrates as salting out agents have been studied by a number of investigations, the extractions from mixed acids into mixed solvents have still to be extracted equilibria.

In the spectrophotometric methods so far developed for the micro determination of uranium in complex materials, though highest sensitivity is provided by the methods developed by Korikisch[15], based on the use of Arsenazo III, they have the disadvantage that most heavy metals, fluorine compounds and phosphate cause errors. Although procedures for diminishing these effects have been suggested, their direct application still yields unsatisfactory results.

Thus the fundamental problem in the precise trace determination of uranium in complex materials is to separate uranium from the matrix components with an extraction system characterised by a high distribution coefficient and good phase separation together with a highly sensitive colour reaction with any organic reagent.

The author has observed that the extraction of V(VI) with dibenzoylmethane (DBM) in presence of masking agents like EDTA or CYDTA can be made more selective quantitative by using a mixed solvent like pyridine in chloroform. The selectively extracted colourless U(VI) - DBM complex can be directly converted into an intensely coloured mixed ligand complex by reaction with 3,5-diBr PADAP in pyridine + chloroform medium which can be made as a highly sensitive method for the precise estimation of uranium. This procedure can also be applied for the highly selective separation of U(VI) of 10-10 mg. from 1 gr. of the matrix components like thorium, iron, copper, molybdenum, vanadium, chromium and zirconium. The data obtained by the author in his systematic and detailed investigations on the solvent extraction of U(VI) and its sensitive and precise estimation as U (VI)-DBM-3, 5-diBr PADAP complex in pyridine + chloroform medium are presented in this chapter.

2.2 Basic Principles of Solvent Extraction

At present solvent extraction is one of the best and most

frequently used methods for the separation of uranium. The method of extraction is extensively used for the separation of elements including uranium, because it is rapid, applicable over a very wide concentration range, and suitable for the separation of the element being analyzed when present in concentrations which are, as a rule, considerably below the solubility of ordinary precipitates. Moreover, by the use of selective extractants and different complex - forming reagents a degree of purification can be attained even with a single-stage extraction, which in the majority of cases is not possible by other methods, especially by the precipitation method. Another advantage of extraction is that it involves no sorption processes, which are characteristic of the precipitation methods and are mainly responsible for lowering the effectiveness of the separation by precipitation. Due to the high specificity of extraction, even the most difficult separations are possible, including the separation of Zirconium from halfnium, niobium from tantalum etc.

The main parameter which characterizes the extraction process for a certain substance is its distribution coefficient, which is the ratio of the concentrations of the substance in the extract and in the extracted solution respectively. D = Corg/Caq. The distribution coefficient remains constant over a wide concentration range, unless the extraction process is complicated by some side effects; in most cases this coefficient is numerically equal to the ratio of the solubilities of the extracted substance in the two phases.

For the separation of two substances, the difference between their distribution coefficients is of fundamental importance. The ratio of the distribution coefficients of two substances $\beta = D1/D2$ is called their separation coefficient; it provides the criterion by which the possible effectiveness of their separation can be judged. The greater the difference between this ratio and unity, the easier is the separation. If $D1 = D2$ the separation becomes impossible.

Besides a high separation coefficient, it is important that the distribution coefficient of the first substance (D_1) should not be too small, otherwise the extractions must be repeated many times with a large volume of the extractant.

Let the initial quantity of the extracted substance in the solution be <u>m</u> and the volume of the solution be <u>v</u>, the quantity of the substance passing into the extract layer m_1 and the volume of the extract V_1. Then the amount of substance which remains in the initial solution after the extraction is $m-m_1$. In this case, the distribution coefficient is

$$D = \frac{m_1/V_1}{(m-m_1)/V} \qquad (1)$$

Hence the amount of substance extracted from the initial solution into the extract can be expressed by means of the following equation

$$m_1 = m \frac{DV_1}{DV_1 + V} \qquad (2)$$

The amount of substance which is extracted as a result of n successive extractions (m_2) is given by the following equation :

$$m_n = m \left(\frac{DV_1}{DV_1 + V} \right)^n \qquad (3)$$

In the separation of a mixture of two substances, each of them is usually extracted independently. Hence for their separation it is necessary that at the moment when one of them has been almost completely extracted, the other substance should remain almost entirely in the initial solution.

By solving equation (3), it is possible to calculate the number of successive extractions which are necessary for the extraction of a definite amount of one of the components of a mixture, as well as the amount of the second substance which will pass into the extract together with the first. For analytical purpose it is often sufficient to achieve the separation of 95-99% of one of the components, the amount of the second component which is extracted along with the first ranges from a few hundredths of one percent to one percent or more.

In the simple cases of separation by extraction, a single extraction achieves a complete separation of the two substances, and the volume of one extractant is considerably less, equal or only slightly larger than that of the initial solution. This is possible when the distribution coefficient of one substance is rather high, while that of the other substance is very low. If the distribution coefficient of one substance is not very high, but that of the other substance is very low, a complete separation can be achieved by a single extraction with a volume of extractant considerably larger than that of the initial solution. In such cases, several successive extractions with small portions of the extractant should be employed instead of a single extraction with a large volume. This device achieves the same result with a much smaller total volume of the extractant.

When the distribution coefficients of both substances are close to one another, i.e., when the separation coefficient is close to unity, the separation becomes a much more difficult task. Yet it is still possible provided that certain conditions are observed. A complete separation of two substances in such cases is achieved by a number of successive extractions. This process is fairly often used on an industrial scale, but very rarely in analytical work.

The distribution law (1) and the fundamental equations for the separation of two components (3) apply when side processes are absent. The chief side effect is the change in the active concentration of the extractable form of the components being separated, which is often not the same as the total concentration of the component. In the case of extraction of electrolytes which are appreciably dissociated into ions in aqueous solutions but which are extracted in the form of undissociated molecules, the concentration of the extractable form (undissociated fraction of the substance) is less than the total concentration of the substance. The distribution law applies to electrolytes if the concentration of the extractable form is used instead of the total concentration of the substance to calculate the distribution coefficient. When the substance is almost completely dissociated into two ions, its concentration in the extract is a linear function of the square of its total concentration in water.

The values of the distribution and of the separation coefficients are very greatly influenced by the presence in the initial solution of other substances including those which tend to combine with water into solvated molecules or ions. It is possible to vary the distribution coefficient over a very wide range by the introduction of neutral salts into the solution which are being extracted; hence this process fairly extensively used.

If the conditions are kept constant, the distribution law holds in every individual case. By means of this law it is possible to calculate the optimum conditions for separation by extraction and to forecast the results fairly accurately.

2.3 Review of Literature on the Solvent Extraction of Uranium

The extraction of uranium appears to have been studied more fully than that of all other elements.

The extractability of the two chief stable valency states of uranium (the tetra-and the hexa-valent) is most pronounced compared to that of the majority of other elements and this is certainly related to the tendency of the UO_2^{2+} and U^{4+} ions to be solvated by the extractant molecules, as well as to their well-developed complex-forming ability with respect to many organic reagents. In the separation of uranium, the extraction of uranium (VI) is by far the most important.

Both uranium (IV) and uranium (VI) (or UO_2^{2+}) have been treated by solvent extraction. Although uranium (VI) forms more stable complexes and thus, in many cases, extracts better than uranium (VI), uranium, has been treated more frequently in the +6 valence state because of its higher chemical stability in air in this state.

Both uranium (IV) and (VI) are class "a" metal ions. They interact very well with those extractants through oxygen and/or nitrogen but very poorly with those through sulfur.

Eversince Peligot found that uranium (VI) in nitric acid is extractable into diethylether[16], the extraction of uranium nitrate with this solvent has been studied by numerous workers[2,17-23]. This extraction without a salting-out agent is not effective, the minimum distribution ratio being only 2.3 in 4 to 5 M nitric acid. Hence quantitative extraction is possible only when nitrate salts are added to the aqueous phase. For example, the extraction is quantitative from 3 M nitric acid containing 2 to 4 M or more concentrated calcium nitrate, or 3 M nitric acid saturated with ammonium nitrate[1,19]

Although diethylether has been used for a very long time by many workers, this solvent in a nitrate system seems to have little value nowadays because of its comparatively low extractability. Complete recovery of uranium (VI) in nitric acid is obtained only after four or more successive extractions. Furthermore it is not very suitable, because of its coextraction of a large amount of nitric acid. Also it has rather high solubility in water. Many other ethers and alcohols have also been studied to ascertain their capacity for extracting uranium from nitrate solutions[20, 22-26]. The extraction with long chain aliphatic ethers is not much better than that of diethylether, but at least these ethers have lower solubility in the aqueous phase[22]. Pentaether (dibutylether of tetraethylene glycol) extracts uranium (VI) from nitrate solutions better than diethylether, but it is inferior to TBP[24].

Among the ketones, MIBK has most frequently been employed for the extraction of uranium in nitrate solution[1, 2, 14, 24, 27-29]. With this reagent the extraction is around 50% over a wide range of nitric acid concentration[24], and the extracted species also contains an amount of nitric acid equivalent to the metal ion.

Trioctylphosphine oxide (TOPO) is also an excellent extractant for uranium in nitrate solution[24,30-37]. From nitric acid, 5% TOPO in toluene extracts uranium quantitatively at acidity levels below about 8.5 M; above this level the extraction falls off. For example, 0.1 M TOPO in cyclohexane extracts uranium quantitatively from 1 to 7 M nitric acid[24,32,34]. The extraction with other trialkylphosphine oxide has also been reported but it is not greatly different from that with TOPO[24,31,35-38]

Other neutral organophosphorus compounds[24,39-41] and alkylsulfoxides[42,43] have also been studied from the standpoint of their capacity to extract uranium (VI) nitrate. However, since not very much merit has been found in the use of these other organophosphorus compounds, TBP and in some TOPO are used by the great majority of workers for this extraction.

Uranium is also extractable with neutral organophosphorus compounds from solutions containing certain other mineral acids or their salts. Uranium (IV) is extracted from hydrochloric acid, in general, better than uranium (VI). The extraction of uranium (IV) and (VI) in hydrochloric acid with TBP has been studied by many workers[11,12,24,44-51]. The extraction of uranium (VI) becomes nearly quantitatively at around 6 M hydrochloric acid when the organic phase is undiluted TBP, but falls off if the acidity is any higher[24]. When TBP is diluted with toluene, the acidity for maximum extraction (where it is practically quantitatively) is somewhat higher. For example, with 50% TBP, the maximum extraction occurs at 8M where the distribution ratio is about 10^2, but with 10% TBP the extraction is no longer quantitative at any acidity[24]. With 30% TBP in carbon tetrachloride, uranium (VI) is effectively extracted from 7 M hydrochloric acid (D=50) and also from 8 M hydrochloric acid (D=102). In the case of uranium (VI), the extraction is also rather effective from 7 M and 8 M hydrochloric acid, the distribution ratios being, respectively, 15 and 21[51].

TOPO is also used for the extraction of uranium from hydrochloric acid[24,33,36,44,49]. The extraction of uranium (VI) with 5% TOPO in toluene is very high: the distribution ratio is above 10^3 when the aqueous phase is 1 to 8 M hydrochloric acid and falls off only to about 10^2 at 10 M, and even at 12 M it is still between 10 and 10^2 [24] uranium in hydrochloric acid is also extractable with other neutral extractants[37,52-54]. However, the extraction with MIBK is only 50% at about 8 M[14,24], and with ethers[24,55] or esters[56] it is much poorer than this.

Uranium in thiocyanate solution can be extracted with ethers[57] alcohols[58], ketones[59], and various oxygen-containing solvents[60,61]. This thiocyante extraction, however, is most often made with TBO[62-64]. The other extraction systems for uranium (VI) which have been reported are in sulfuric acid with TBP[24], and with TOPO[65] and others[38,53,66] in perchloric acid with ethers[55]. TBP[67-69] and dialkylsulfoxides[70] in trichloroacetic acid with ethers, alcohols, and other solvents[23,71] and chloroacetate acid with ethers, alcohols, and other solvents [23,71] and in solutions containing an organic ligand with TBP[72]. Extraction of uranium (VI) from alkaline solutions with TBP has also been studied[73].

The amines commonly used for the extraction of uranium (VI) from nitrate solutions are trioctyl - and triisooctylamine[24,74,75]. trilaurylamine[76], Alamine 336[77], and other trialkylamines[78], but various other amines[24,79,80] and amine oxides [81] have also been studied on their extraction of uranium (VI) nitrate.

Uranium (VI) complexes can be extracted as ion-pairs with large cations. Nitrate complexes are extractable with quaternary ammonium salts[82,83,86] and others[84]. chloride[85], bromide[86], thiosulfate[87] and carbonate[88] complexes are also extractable with quaternary ammonium salts. There have also been reports on the extraction of uranium (VI) complexes with benzoate ion and with other ligands with quaternary ammonium salts or dyes[89-91].

The extraction of uranium with various alkylphosphoric acids has been studied by many workers. The extractants of this type that are often used are dibutylphosphoric acid (DBP)[40,92] mixtures of DBP and TBP or TOPO[93] and dioctylphosphoric acid[94].

Among the alkylphosphoric acids, however, di (2-ethylhexyl) phosphoric acid (DEHP) is most frequently used for the extraction of uranium[92, 50-101]. In general, the extraction with alkylphosphoric acid is better at higher pH. When DEHP is used without dilution, the extraction of uranium (VI) from 4 M nitric acid is 66%, while that from 2M hydrochloric acid is 96%[102]. When 50% DEHP in toluene is used, the extraction is quantitative from aqueous solution containing hydrochloric acid at 1 M or less[24]. The extraction improves as pH goes higher and diminishes as the concentration of the extractant becomes lower, but these changes may deviate from those statistically expected, especially when the pH is low or the concentration of the reagent is low. This reagent was also used for the extraction of uranium (III) [92] and uranium (IV) [98,101]. There are many papers on the extraction of uranium (VI) with DEHP in various solvents in the presence and absence of adduct forming ligands, especially of TBP and TOPO[95,103].

Both uranium (IV) and uranium (VI) are extractable with various chelating extractants. Since uranium (VI) forms stable complexes with carbonbate ions in aqueous solution, some of the equilibrium constants for the chelating extraction of uranium (VI) carried out in air could be different from the values obtained in carbonate-free systems due to this complex formation.

B-Diketones are very effective for the extraction of uranium. It is interesting that uranium (VI) is extracted with a B-diketone (HA) often as a self-adduct, $UO_2A_2(HA)$. With 0.5 M acetlylacetone in benzene or in chloroform, the extraction of uranium (VI) is quantitative above pH 3[104] and that of uranium (VI) with pure acetylacetone is more than 95% at pH 2 to 7[105]. With acetylacetone in chloroform or benzene, uranium (VI) is extracted as both UO_2A_2 and $UO_2A_2(HA)$[106-108], but the extraction is not complete : with 0.1 M acetylacetone, the extraction is about 50% above pH 5 : and with 1 M acetylacetone, the maximum extraction is 90% above pH 4[107]. The absorption spectra of the complex have been presented[109]. There are also some other reports on uranium extraction with acetylacetone[110,111] and its derivatives[112].

Benzoylacetone and dibenzoylmethane are also effective extractants for uranium (VI)[28,32,111,113-118]. The extraction of uranium (VI) with 0.1 M benzoylacetone in benzene, chloroform or carbon tetrachloride is quantitative at pH 5 to 7, and the extracted chelate complex, identified to be in the form $UO_2A_2(HA)$, absorbs at 380 nm[116]. The extraction of uranium (VI) with 0.01 M dibenzoylmethane in benzene, chloroform, or carbon tetrachloride is quantitative at pH 5 to 6.5[118] and the extracted species, $UO_2A_2(HA)$, absorbs at 400 mm ($E = 2 \times 10^4$)[114,117].

In addition to the above reagents, various B-diketones and pyrazolones have also been used for the extraction of uranium[111,119,120]. However, the B-diketone used most frequently is TTA[111,121-124]. It has been reported that uranium (IV) is extractable with TTA from acidic solutions[122], but the extraction of uranium (VI) with TTA can be accomplished at somewhat higher pH: with 0.15 M TTA in benzene, it is quantitative at pH 3.5 to 8[123]. The extracted species are UO_2A_2 and $UO_2A_2(HA)$[106 124], Synergic enhancement of the extraction of uranium (VI) with TTA and other B-diketones has also been studied by many workers[111,121,125-127].

The extraction of uranium (VI) with 0.1 M B-isopropyltropolone (HA) in chloroform was found to be quantitative above pH 1, and the extracted species was identified to be $UO_2A_2(HA)$[128].

Oxine has frequently been used for the extraction of uranium[129-134] The extraction of uranium (VI) with 0.1M oxine in chloroform was quantitative at pH 5 to 9, and the extracted species was reported to be $UO_2A_2(HA)$[133]. The complex absorbed at 425 to 430 nm[132,134]. 5, 7-Dihalogen derivatives of oxine were reported to extract uranium at somewhat lower pH than oxine[132, 135].

1-Nitroso-2-naphthol forms a chelate with uranium (VI)[136]. Extraction of this complex incomplete when the organic phase is chloroform. It is quantitative when the organic phase is butyl alcohol or ethylacetate, and the pH in the aqueous phase is 3.0 to 8.5; the pH region of complete extraction is narrower when the organic phase is isoamyl alcohol[136].

Uranium (VI) forms a complex with cupferron which is extractable from acid solution into diethylether or ethylacetate[137,138]. The extraction of uranium (VI) with cupferron, however, is poorer than that of uranium (VI) : with 0.01 M cupferron in chloroform, the extraction is less than 30% at pH 3.5 to 6, and at lower pH and cupferron concentrations, the extraction is negligible[139] although the extraction is poor, the nature of uranium (VI) cupferrate has been studied[140]. By making use of the difference between the extraction of uranium (VI) and uranium (VI) cupferrates, uranium in these two valence states can be mutually separated[138]. The extraction uranium (VI), with N-benzoyl-N-Phenylhydroxylmine and its homologues has been reported[137, 141, 142]. For example, 0.1 M of this reagent in chloroform gives more than 90% extraction when the pH is higher than 3.5[141]. Hydroxamic acids have also been used for the extraction of uranium (VI)[143,144]. The extraction of uranium with PAN[145,146] and its homologue has also been reported by several workers. For example, the extraction of uranium (VI) into chloroform is possible when the aqueous phase contains 0.1% PAN and the pH is 5 to 10. and the complex formed absorbs at 560 nm[145]. The complex is also extractable into-o-dichlorobenzene, and the complex in this solvent absorbs at 570 nm (e = 2.3 x 10^4)[146]. Thio-TTA extracts uranium (VI) at pH 5.5 to 6 and the complex absorbs at 490 nm[147].

Diethyldithiocarbamate in various solvents is another useful extractant for uranium (VI)[148, 149]. Although the extraction with this reagent into carbon tetrachloride is poor, it is good with other solvents such as benzene, chloroform, alcohols, and ketones. The absorption spectrum of the complex is observed at 300 to 620 nm[149]. The extraction of uranium (VI) with 0.04% diethylammonium diethyldithiocarbamate in chloroform is quantitative at pH 6.5 to 8, but when the solvent is carbon tetrachloride it is very poor[150]. At pH 5.5 to 6.5% of this reagent in chloroform is reported to be effective for uranium[150] Pyrrolidinedithiocarbamate[151], xanthate[148] and various miscellaneous extractants[152] have been used for the separation of uranium.

Unlike other metals uranium (VI) does not form complexes with EDTA and CYDTA which can be used as masking agents for decreasing the interference of co-existing metal ions. If EDTA is used along with magnesium chloride it enhances the extraction kinetics of uranium (VI) (facilitating for a fast selective separation of uranium from co-existing matrix components.

At pH 6.5 - 8.5 the enol form of dibenzoylmethane forms a yellow uranyl chelate complex UO_2R_2, which can be used for Extractive spectrophotometric determination of uranium.

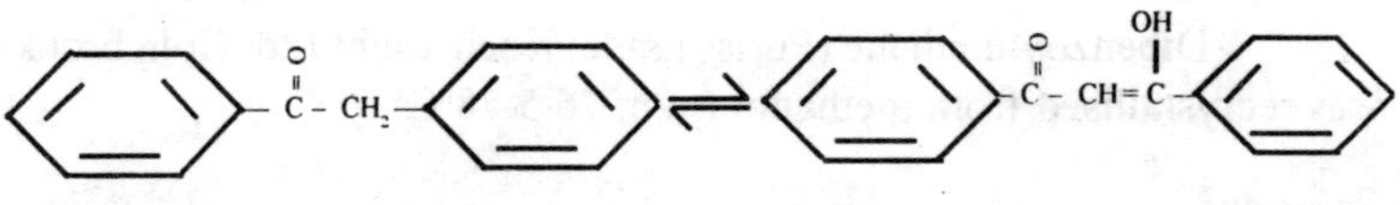

The absorption maximum of the complex in ethyl acetate is at 395 nm, molar absorptivity 2.0 x 10^4 (a=0.088) dibenzoylmethane itself absorbs at wavelength shorter than 400 nm. Phosphate, citrate and most metals forming dibenzoylmethane complexes interfere. The uranium (VI) - chelates of dibenzoylmethane can be extracted by many water-immisicble organic solvents like chloroform, carbontetrachloride, benzene, toluene, xylene, etc. which give coloured extracts.

Though EDTA, CYDTA can be used as good masking agents for suppressing the interference of several metal ions, in presence of these masking agents, the extraction of uranium becomes less quantitative requiring several extractions with increased equilibration periods.

However, it is observed by the author that when mixed solvent like pyridine in chloroform or pyridine in carbon tetrachloride is used, the extraction becomes quantitative in a single extraction over a wider pH range and in presence of calcium nitrate + aluminium nitrate + EDTA the extraction becomes selective and quantitative. Further it is also observed by the author that the selectively extracted colourless uranium (VI) - DBM complex can be converted by reacting with a chelating chromogenic complexing agent like 3,5 diBr PADAP in a surfactant medium like pyridine in chloroform. Due to interaction of dye and surfactant in micellar medium it is observed that the colour of uranium (VI) - DBM - 3,5, diBr PADAP system is sensitized further by increasing the colour sensitivity accompanied with red-shift in absorption band. The advantageous effect of micellar system on colour formation of uranium (VI)-DBM-3,5 diBr PADAP complex has been exploited for developing a highly selective, precise, sensitive and rapid estimation of uranium (VI). The author's detailed experimental results on the solvent extraction of U(VI)-DBM under varied experimental conditions and the spectrophotometric studies on the U(VI)-DBM-3,5-diBr PADAP in non-aqueous medium are presented in the following pages :

EXPERIMENTAL

2.4 Reagents and Apparatus

Dibenzoylmethane (Puriss grade; Koch-Light Ltd; Coln brook) was recrystallised from methanol (m.p. 76.5-78°C).

Solutions

1. *Dibenzoylmethane solution* : - A 0.2% (w/v) solution of dibenzoylmethane was prepared by dissolving 2.0 gms in 700 ml chloroform, adding 20.0 ml of tri-n-octylamine (TOA), and diluting to one liter with chloroform.
2. *Standard uranium solution* : A stock solution, containing 2.0 mg. U/ml was prepared by dissolving 1.179 gms of U_3O_8 in 20 ml of

concentrated hydrochloric acid and 3 ml of concentrated nitric acid. The resulting solution was evaporated to dryness. Then 5 ml of hydrochloric acid (1+1) was added, and the solution was again evaporated to dryness. 20 ml of hydrochloric acid (1+1) was added, and the solution was diluted to exactly 500 ml with water. A working standard solution of 10-100 μg/ml was prepared by diluting this stock solution with water.

3. *3,5 diBr PADAP Solution* : 0.02 gm of 3,5-dibromo-PADAP (Fluka) is dissolved in 100 ml absolute ethanol.
4. *Aluminium nitrate solution* : 25 gms of aluminium nitrate Al $(No_3)_3$ $9H_2O$ was dissolved in 500 ml of distilled water by heating.
5. *Calcium nitrate solution* : a 1% solution of calcium nitrate was prepared by dissolving 2.5 gms in 250 ml. water.
6. *Tetrasodium EDTA solution* : 25 gms of disodium EDTA was dissolved by slowly adding to it 134 ml. of 1.0 M sodium hydroxide solution, and then diluting to 250 ml with distilled water.
7. *Sodium fluoride solution* : 5.0 gms of sodium fluoride was dissolved in 250 ml volumetric flask and made up to the volume with distilled water.

Equipment

The absorbance measurements were made on a Shimadzu U.V-visible spectrophotometer. The pH measurements were made using Systronics model 108 pH meter.

2.5 General Extraction Procedure

To 5 ml. of slightly acidic solution of uranium (VI) (200 μg.), 0.5 ml each of the following masking agent solutions (15% EDTA and 2% sodium fluoride) are added and allowed to stand for 2 minutes. To this, 2.5 ml each of 1% calcium nitrate and 5% aluminium nitrate solutions are added. The total volume is made upto 25 ml. adjusting the pH to required value by adding enough dilute sodium hydroxide solution or hydrochloric acid. To this 5 ml of 0.2% dibenzoylmethane in chloroform and 20 ml (15% pyridine + 10% n butanol in chloroform) mixed solvent are added and equilibrated for 2 minutes. After clear separation of the phases, the uranium content in both organic and aqueous phases are estimated by adopting uranium (VI) DBM-3.5-diBr PADAP spectrophotometric method described in 3.3.

Table - 2.1
Effect of The Nature of Solvent on the Extraction of U U(VI)-DBM-System

Aqueous phase 5.0 ml of U (VI) (40 μg/ml) + 0.5 ml each of 15% EDTA and 2% sodium fluoride masking agent solution + 2.5 ml each of 1% calcium nitrate and 5% aluminium nitrate salting-out agent solutions, total volume 25 ml, pH : 6.

Organic phase 5.0 ml of 0.2% dibenzoylmethane in chloroform + 20 ml of various solvents as diluents for 15% pyridine and 10% n-butanol, total volume 25 ml.

S. No.	*Solvent*	*% E in Single Extraction*	*Remarks*
1	2	3	4
I.			
1.	Carbon tetrachloride	80.0	Slow attainment
2.	15% pyridine + carbon tetrachloride	99.0	of Equilibrium.
3.	10% n-butanol + carbon tetrachloride	95.0	
4.	15% pyridine + 10% n-butanol + carbon-tetrachloride	99.0	
II.			
1.	Chloroform	92.5	Easy separation
2.	15% pyridine + chloroform	95.9	of two phases &
3.	10% n-butanol + chloroform	96.0	Equilibrium
4.	15% pyridine + 10% n-butanol chloroform	95.9	achieved quickly.
III.			
1.	Benzene	75.0	Easy separation
2.	15% pyridine + benzene	95.0	of two phases %
3.	10% n-butanol + benzene	89.0	Equilibrium
4.	15% pyridine + 10% n-butanol + benzene	96.6	achieved quickly.
IV.			
1.	Nitrobenzene	90.5	Organic extract
2.	15% pyridine + nitrobenzene	95.0	turbid
3.	10% n-butanol + nitrobenzene	95.0	
4.	10% n-butanol + 15% pyridine + nitrobenzene	98.0	

(Contd.)

Table - 2.1 (Contd.)

1	2	3	4
V.			
1.	Methyl iso butyl ketone	99.0	Organic extract
2.	15% pyridine + MIBK	99.0	turbid
3.	10% n-butanol + MIBK	95.0	
4.	10% n-butanol + 15% pyridine + MIBK	99.0	
VI.			
1.	Ethyl-acetate	85.0	Organic extract turbid
VII.			
1.	Isoamyl alcohol	90.0	Organic extract turbid

2.6 Effect of Nature of Solvent on the Extraction of U(VI)-DBM System

In the extraction of uranium as the neutral donors like pyridine and n-butanol were found to exert synergistic effect fácilitating quantitative extraction of uranium. The extraction efficiency of various extracting solvents like carbon tetrachloride, chloroform, benzene, MIBK, ethylacetate, isoamyl alcohol in the presence and absence of pyridine and n-butanol has been systematically investigated and the results are presented in Table - 2.1.

Though quantitative extraction can be achieved with carbontetrachloride, chloroform, benzene, nitrobenzene, MIBK in the presence of pyridine and n-butanol, 15% pyridine + 10% n-butanol in chloroform is found to be the best extracting mixed solvent for the extraction of U(VI)-DBM system keeping in view of the speedy attainment of equilibrium with intense colour development with 3.5-diBr-PADAP system.

2.7 Effect of Concentration of Pyridine and N-Butanol

The effect of concentration of pyridine and n-butanol on the extraction of U (VI)-DBM system with pyridine in chloroform is investigated by following the general extraction procedure with different concentrations of n-butanol and pyridine in chloroform and the corresponding results are included in Table 2.2.

Though there is a marginal increase in the extraction of U(VI)-DBM in presence pyridine + n-butanol it is observed that pyridine + n-butanol medium facilitates quantitative extraction of U(VI) in a single extraction. Further, the prescnce of pyridine + n-butanol is found to be essential for the colour reaction between U(VI)-DBM and chromogenic chelating agent 3,5-diBr-PADAP. So, 15% pyridine + 10% n-butanol in chloroform is used in all further investigations.

2.8 Effects of pH on the Extraction of U(VI)-DBM System

Uranium (VI) known to be extracted in the pH 1-10 with various solvents like carbontetrachloride, chloroform and requires 2-3 extractions with fresh solvent for quantitative extraction, particularly in presence of masking agents. To understand the effect of pH on the extraction of U(VI)-DBM with 15% pyridine + 10% n-butanol in chloroform mixed solvent in presence and absence of masking agents, a systematic study has been taken up.

Table 2.2

Effect of Concentration of Pyridine on the Extraction of U(VI)-DBM System with Pyridine and n-Butanol in Chloroform

Aqueous phase : 5.0 ml of U (VI) (40 μg/ml) + 0.5 ml each of 15% EDTA and 2% sodium fluoride masking agent solution + 2.5 ml each of 1% calcium nitrate and 5% aluminium nitrate salting-out agent solutions, total volume 25 ml, pH : 6.

Organic phase : 5.0 ml of 0.2% dibenzoylmethane in chloroform + x ml of pyridine and n-butanol in chloroform, total volume 25 ml.

S. No.	*% n-butanol*	*% Pyridine*	*% E*
1.	10	1	90.0
2.	10	3	94.5
3.	10	5	95.0
4.	10	7	95.0
5.	10	10	99.9
6.	5	15	99.9
7.	5	15	95.0
8	10	20	95.9
9.	10	20	99.0
10.	10	30	99.0
11.	10	40	99.0
12.	10	50	99.0

pH values of the aqueous phase from 0.1 - 9.5 were adjusted by adding either sodium hydroxide or hydrochloric acid, to obtain desired pH value, using a pH meter.

In the study, in presence of masking and salting-out agents, extractions were carried out in presence of 0.5 ml each of 15% EDTA and 2% sodium fluoride and 2.5 ml each of 1% calcium nitrate and 5% aluminium nitrate and the data is included in the Table 2.3.

Table 2.3

Effect of pH on the Extraction of U(VI)-DBM System

Aqueous Phase : 5.0 ml of U(VI) 40 μg/ml+X ml of acid or alkali solution for required pH total volume 25.0 ml.

In presence of masking agents :

5.0 ml of U(VI) 40 μg/ml+0.5 ml each of masking agent solutions 15% EDTA and 2% sodium fluoride + x ml of acid or alkali solution for required pH.

Organic Phase : 5.0 ml of 0.2% DBM in chloroform + 20 ml of (15% pyridine + 10% n-butanol in chloroform) mixed solvent. Total Volume 25 ml.

S. No.	*pH*	*% E in the absence of masking agent*	*% E in presence of masking agent*		
			EDTA	*sodium fluoride*	*EDTA + fluoride*
1.	0.1	15.0	15.0	15.0	15.0
2.	0.5	32.0	25.0	22.0	22.0
3.	1.0	45.0	45.0	45.0	45.0
4.	2.0	50.0	55.0	55.0	55.0
5.	3.0	55.0	80.0	60.0	60.0
6.	4.0	65.0	98.0	98.0	98.0
7.	5.0	75.0	98.0	98.0	98.0
8.	6.0	85.0	98.0	98.0	98.0
9.	7.0	80.0	98.0	98.0	98.0
10.	8.0	75.0	75.0	75.0	80.0
11.	9.0	55.0	55.0	55.0	65.0
12.	9.0	45.0	45.0	45.0	55.0

It is well known that while at pH values less than 2.5 uranium (VI) exists as uranyl ion (UO_2^{2+}), at pH values higher than 2.5, it undergoes hydrolysis forming predominate complex hydrolysed ions which hinder the extraction of uranium. However, DBM undergoes protonation, while its neutral species which exist in between 2.5 - 7.0 pH form extractable neutral chelates with uranium. In view of the above, quantitative extraction of uranium with dibenzoylmethane cannot be achieved in the absence of any masking mixture, which can be observed from the data given in the Table 2.3. In the presence of masking mixture like EDTA and sodium fluoride, which form weak complexes with uranium and act as auxiliary ligands preventing the hydrolysis of uranium and the extraction of uranium in their presence becomes quantitative with dibenzoylmethane. However, above pH 7 as stronger composite hydrolysed ions of uranium predominate resulting in its precipitation, the extraction steeply falls, both in the presence and absence of masking agents. Thus, in the extraction of uranium, use of masking agent mixture like EDTA and sodium fluoride will be useful both for preventing hydrolysis of uranium between 2.5 - 7 pH and for suppressing the interference of other associated metal ions.

2.9 Effect of Volume of Aqueous Phase on the Extraction of U(VI)-DBM System

To examine the influence of the volume of aqueous phase on the extraction efficiency of U(VI)-DBM pyridine + n-butanol in chloroform system, 200 μg. of U(VI) is extracted from 20, 50, 100, 200 ml of total volume of the aqueous phase into 25 ml of 15% pyridine + 10% n-butanol in chloroform. Quantitative extraction of U(VI) is achieved in all the cases. However, with 100 and 200 ml volume of the aqueous phase larger equilibration periods (5 minutes) are found to be necessary.

2.10 Effect of Shaking Time on the Extraction of U(VI)-DBM System

200 μg of U(VI) is extracted into 15% pyridine + 10% n-butanol in chloroform mixed solvent adopting the general extraction procedure with different shaking periods. It is observed that a two minute shaking period is sufficient for attaining quantitative extraction of uranium (VI)-DBM with 15% pyridine + 10% n-butanol in chloroform. It is further observed that in a second extraction carried out no significant amount of uranium (VI) is observed, which reveals that the first extraction itself

removes quantitatively uranium as U(VI)-DBM into 15% pyridine + 10% n-butanol in chloroform.

2.11 Effect of Concentration of Dibenzoylmethane on the Extraction of U(VI)-DBM System into 15% Pyridine + 10% N-Butanol in Chloroform Mixed Solvent

The effect of DBM on the extraction of U(VI)-DBM system with 15% pyridine + 10% n-butanol in chloroform mixed solvent is investigated varying the concentration of DBM by adopting the general extraction procedure and corresponding results are given in Table 2.4.

TABLE 2.4

Effect of Concentration of DBM on the Extraction of U(VI)-DBM System with 15% Pyridine + 10% n-Butanol in Chloroform mixed Solvent

Aqueous Phase : 100 μg/200μg/500μg/1000 μg U(VI) - 0.5 ml each of 15% EDTA and 2% sodium fluoride + x.0 ml of $10^{-3}/10^{-2}$ M DBM (0.4×10^{-4}; 4×10^{-4} M in 25 ml) : total volume 25.0 ml.

Organic Phase : 25.0 ml 15% pyridine + 10% n-butanol in chloroform mixed solvent

S. No.	*DBM x 10^{-4} M*	*Initial Aqueous Concentration of U (VI)*			
		100 μg	*200 μg*	*500 μg*	*1000 μg*
		% E U (VI)			
1.	0.4000	90.0	75.0	50.0	30.0
2.	0.8000	98.0	80.0	60.0	40.0
3.	1.2000	99.9	85.0	75.0	50.0
4.	1.6000	99.9	90.0	80.0	60.0
5.	2.0000	99.9	98.0	85.0	70.0
6.	2.4000	99.9	99.9	90.0	75.0
7.	2.8000	99.9	99.9	95.0	80.0
8.	3.6000	99.9	99.9	99.9	85.0
9.	6.0000	99.9	99.9	99.9	90.0
10.	8.0000	99.9	99.9	99.9	98.0
11.	12.0000	99.9	99.9	99.9	99.9
12.	16.0000	99.9	99.9	99.9	99.9

2.12 Interference Studies

As EDTA and fluoride form very weak auxilary complexes with uranium (VI) when compared with dibenzoylmethane, they can be tolerated in large concentration in the extraction of uranium with dibenzoylmethane in the pH range 4 to 7. Further they prevent the hydrolysis of uranium and mask the interference of many other metal ions which form strong colourless complexes with EDTA and fluoride. Further, any excess of these masking agents can be removed by the addition of aluminium nitrate and calcium nitrate as salting out agents.

The sulphate ion in amounts upto 2 mg/ml has no appreciable effect on the completeness of the extraction of uranium. If it is present in large amounts, its interference can be removed by using calcium nitrate as the salting-out agent. If phosphates and sulphates are present simultaneously, a salting out agent which consists of a mixture of aluminium and calcium or iron and calcium nitrates can suppress their interference upto 0.5 mg.

So, the interference of other anions and cations on the extraction of U(VI)-DBM-pyridine system has been studied in presence of EDTA and sodium fluoride as masking agents and calcium nitrate and aluminium nitrate as salting out agents. These results are presented in Tables 2.5 and 2.6.

Table 2.5

Effect of Diverse Ions of the Extraction of U(VI)-DBM System in 15% Pyridine + 10% n-Butanol in Chloroform Mixed Solvent

Aqueous Phase : 5.0 ml of 80 μg/ml U(VI) + X ml of diverse ion + 0.5 ml each of 15% EDTA and 2% sodium fluoride + 2.5 ml each of 1% calcium nitrate and 5% aluminium nitrate, total volume 25 ml, pH : 6.

Organic Phase : 5.0 ml of 0.2% DBM in chloroform + 20 ml of (15% pyridine + 10% n-butanol in chloroform) mixed solvent.

S. No.	*Foreign Ion/ neutral ligand*	*Added as*	*Amount of ion added*	*Tolerance limit for achieving quantitative extn of U(VI) (μg)*
1	2	3	4	5
1.	Cl^-	KCL	50 mg	50000
2.	Br^-	NH_4Br	50 mg	50000
3.	I^-	KI	50 mg	50000
4.	NO_3^-	KNO_3	50 mg	50000
5.	SO_4^{2-}	$Na_2SO_45H_2O$	50 mg	50000 > 1.0 mg
6.	PO_4^{3-}	Na_3PO_4	50 mg	interfere
7.	$Ascorb^-$	Ascorbic acid ($C_3H_7O_6$)	50 mg	50000

(Contd.)

Table 2.5 (Contd.)

1	2	3	4	5
8.	Carbonate	Na_2CO_3	50 mg	interfere > 1.0 mg
9.	$Tart^{3-}$	Tartaric acid	50 mg	interfere > 1.0 mg
10.	Cit^{3-}	Citric acid	50 mg	interfere > 1.0 mg
11.	Monochloro acetate	Monochloro acetic acid	50 mg	50000

Table 2.6

Effect of Diverse Ions on the Extraction of U(VI)-DBM System in 15% Pyridine + 10% N-Butanol in Chloroform Mixed Solvent

S. No.	*Foreign cation*	*Added as*	*Amount of cation added (µg)*	*Tolerance limit for achieving quantitative extn of U(VI) (µg)*
1.	Hg^{2+}	$Hgcl_2$	1000	Coextracts
2.	Pb^{2+}	$Pb(NO_3)_2$	1000	1000
3.	$Bacl_2$	$Bacl_2$	1000	1000
4.	Cu^{2+}	$CuSO_4 5H_2O$	1000	1000
5.	Cd^{2+}	$3CdSO_4 8H_2O$	1000	1000
6.	Sb^{3+}	$Sb_2(SO_4)_3$	1000	1000
7.	Sn^{4+}	$SnCL_4$	1000	1000
8.	Pt^{4+}	H_2PtCl_6	1000	1000
9.	Fe^{3+}	$FeCl_3 6H_2O$	1000	1000
10.	Ni^{2+}	$NiSO_4 7H_2O$	!000	1000
11.	Co^{2+}	$CoSO_4 7H_2O$	1000	1000
12.	Mn^{2+}	$MnSO_4 7H_2O$	1000	1000
13.	V^{5+}	NH_4VO_3	1000	1000
14.	Mg^{2+}	$MgSO_4 7H_2O$	1000	1000
15.	Ti^{4+}	·Titanium Sponge	1000	Huge ppt.
16.	Ca^{2+}	$CaCl_2.2H_2O$	1000	1000
17.	Al^{3+}	$Al(NO_3)_3.9H_2O$	1000	1000
18.	Cr^{3+}	K_2Cro_4	1000	1000
19.	Nb^{5+}	Nb_2O_5	1000	1000
20.	W^{6+}	Na_2WO_4	1000	Interfere
21.	Zn^{2+}	$Zn(NO_3)_2$	1000	1000
22.	Zr^{4+}	$Zr(NO_3)_4 5H_2O$	1000	Interfere
23.	Th^{4+}	$ThNO_3$	1000	Interfere

Cd, Za, Cr, Mg, Ba and Sb do not interfere even in the absence of masking mixture while the masking mixture of EDTA and sodium fluoride binds Hg, Pb, Cu, Sn, Pt, Fe, Ni, Co, Mn, Nb, Zn and remove their interference in the extraction of uranium (VI) with U(VI)-DBM-pyridine system. However, serious interference is caused by Mo, W, and Th arsenates, Carbonates, Phosphates, Tartrates and Citrates.

2.13 Separation Factor

The term separation factor may be used as an index of effectiveness of metal ions which is one of the main objectives of solvent extraction.

The relationship between extraction efficiency E and distribution coefficient is

$$E = 100D/ \{ D + (V_w/V_o) \} \quad (4)$$

Where V_o, V_w are the volumes of the organic and aqueous phases.

The ease of separation of two metals, both of which form extractable chelates with a particular reagent, may be readily given by the separation factor. β

$$\beta = \frac{D_1}{D_2} = \frac{Kf_1 K_{DX1}}{Kf_1 K_{DX2}} \quad (5)$$

The subscripts 1 and 2 denote the metals in question, D is the ratio of the stoichiometric metal concentration in the two phases. K_f is the overall formation constant for the chelate and k_{DX} is equilibrium constant for the distribution of the chelate.

If we accept a minimum of 99% extraction of metal 1 with a corresponding maximum of 1% extraction of metal 2, then as

$$\beta = \frac{D_1}{D_2} = \frac{10^2}{10^{-2}} = 10^4 \quad (6)$$

the $K_f K_{DX}$ product for metal 1 must be at least 10,000 times greater than that of metal 2 for separation with a one-stage extraction. Separations of practical value have been carried out in two or three extraction stages where the distribution ratios of the two metals differed by much smaller values.

In fact, if extraction of more than 99% of one metal and co-extraction of less than 1% of another metal is desired by a single extraction process, then a separation factor higher than 10^4 is necessary. The extraction of U(VI) with DBM using 15% pyridine + 10% n-butanol in chloroform in presence of masking agent mixture and certain metal ions which are likely to encounter in complex matrices in varied concentrations, has been examined. To understand the selectiveness and the practical applications of the extraction, separation factors are also calculated.

Procedure for the Determination of Separation Factor

Transfer 5.0 ml of 40 μg/ml uranium solution, x.0 ml of other diverse ions, 0.5 ml each of 15% EDTA and 2% sodium fluoride and allow to stand for 2 minutes. To this, add 2.5 ml each of 1% calcium nitrate and 5% aluminium nitrate. Make up the volume to 25 ml maintaining the pH to 6.5 by adding dilute sodium hydroxide or hydrochloric acid. To this, add 5 ml of 0.2% dibenzoylmethane in chloroform and 20.0 ml of 15% pyridine + 10% n-butanol in chloroform. Equilibrate the mixture for 2 minutes. After clear separation of the two phases, remove the organic phase and back extract uranium with 25.0 ml of 2N HNO_3. Then determine the uranium present in the original aqueous phase and in the back extract of the organic extract by U.V. Spectrophometer. The amounts of diverse ions present in the aqueous phase after separation from U(VI) are determined by corresponding standard procedures. The results obtained in these investigations are given in Table 2.7.

It can be seen from the data given in Table 2.7 that very effective separation of U(VI) from a number of diverse metal ions can be achieved with DBM into 15% pyridine + 10% n-butanol in chloroform in presence of masking agent mixture.

2.14 Nature of Extracted Species

Neutral extractants like pyridine often extract uncharged metal complexes in aqueous solution, and under certain conditions they also extract charged metal ions and complexes.

When a metal ion M^{m+} in an aqueous solution undergoes complex formation with an anionic ligand L^- (more generaly L^{l-}), metal species, M^{m+}, ML^{m-l},...... ML_m,...., ML^{m-n}_n are present in the solution. Among these chemical species, the uncharged complex, ML_m, enters the organic phase usually more easily than the others.

Table 2.7
Selective Separation of U(VI) from Certain Diversified Metal Ions

Aqueous Phase : 5.0 ml of U(VI) (40 #g/ml) + x.0 ml solution of other diverse metal ion + 0.5 ml each of 15% EDTA and 2% sodium fluoride masking agent solutions + 2.5 ml each of 1% calcium nitrate and 5% aluminium nitrate salting out agent solutions, total volume 25 ml, pH : 6.

Organic Phase : 5 ml 0.2% DBM in chloroform + 20 ml of (15% pyridine + 10% n-butanol in chloroform) mixed solvent:

S. No.	*Composition µg*	*Recovery of U(VI)*	*Recovery of added ion %*	*Separation factor*
1.	U(VI) : Ni (II) 200 : 500	97.0	96.5	0.12×10^4
2.	U(VI) : Mn (II) 200 : 1000	97.0	96.0	0.10×10^4
3.	U(VI) : Zn (II) 200 : 800	95.0	96.0	0.20×10^4
4.	U(VI) : MG (II) 200 : 5000	98.0	96.0	0.20×10^4
5.	U(VI) : Co (II) 200 : 500	96.0	97.5	0.20×10^4
6.	U(VI) : Sb (II) 200 : 600	99.0	97.0	0.12×10^4
7.	U(VI) : Cu (II) 200 : 500	97.0	96.5	0.12×10^4
8.	U(VI) : Sn (IV) 200 : 600	96.0	97.0	0.95×10^3
9.	U(VI) : Cr (III) 200 : 1000	96.0	95.5	0.16×10^4
10.	U(VI) : Cd (II) 200 : 800	98.0	97.0	0.20×10^4
11.	U(VI) : Fe (III) 200 : 1000	96.0	95.5	0.20×10^4
12.	U(VI) : Pb (II) 200 : 500	97.0	96.5	0.12×10^4

The extraction of most complexes of this type into unsolvating solvents, however, is too poor to determine by ordinary experimental methods. An addition of an organophilic uncharged ligand having the capability of farming stable adduct complexes with the ML_m species is necessary. Reagents such as pyridine bases, esters, ketones, or alcohols are used for this purpose without dilution. Other reagents, such as organophosphorus compounds, are often dissolved in a nonpolar solvent, called an inert diluent, examples of which are hydrocarbons and carbon tetrachloride.

The complex formation of M^{m+} with an anionic ligand L^- in aqueous solutions can be described as follows :

$$M^{m+} + nL^- \rightleftharpoons ML_n^{m-n}$$

$$Bn = \frac{ML_n^{m-n}}{[M^{m+}][L^-]^n} \quad \text{——— (1)}$$

When only the uncharged complex is extractable into an inert solvent containing on extractant, the distribution equilibrium is expressed as :

$$ML_m \rightleftharpoons ML_m(org) \quad \text{——— (2)}$$

$$K_{dmo} = \frac{[ML_m]_{org}}{[ML_m]} \quad \text{——— (3)}$$

If the extracted species forms adduct complexes with an organophilic neutral extractant, E, the equilibrium in the organic phase is :

$$ML_m(org) + eE\ (org) \rightleftharpoons ML_mE_{e\,(org)} \quad \text{——— (4)}$$

$$\beta_{e\,(org)} = \frac{[ML_m\ E_e]_{org}}{[ML_m]_{org}\ [E]^e_{org}} \quad \text{——— (5)}$$

The distribution ratio of M can then be rewritten from the above equations as :

$$D = \frac{[ML_m]\ org + [ML_mE]\ org + \text{----} + [ML_m\ E_e]\ org}{[M^{m+}] + [ML^{m-1}] + \text{---} + [ML_m^{-1}] + \text{---} + [ML_n^{m-n}]} \quad \text{—— (6)}$$

$$D = \frac{Kdmo\ \beta_m\ [L^-]^m\ (1 + \Sigma\ \beta_e\ (org)[E]^e_{org})}{1 + \Sigma\ \beta_n\ [L^-]^n} \quad \text{—— (7)}$$

When a series of distribution experiments are carried out by using an undiluted extractant such as an ether, ketone, or alcohol, or by keeping the concentration of the extractant at a certain value, the concentration of the extractant, $(E)_{org}$, can be eliminated from the equation and the distribution ratio can be expressed as :

$$D = \frac{K_{dm}\beta_m[L^-]^m}{1+\Sigma\beta_n[L^-]^n} \quad \text{———} \quad (8)$$

$$K_{dm} = K_{dmo}(1+\Sigma\beta_e(org)[E]^e_{org}) \quad \text{———} \quad (9)$$

Uranium(VI) is present in solutions as Uo_2^{2+} ion, the existence of which at pH ≤ 2.5 has been firmly established [153,154]. At higher pH values composite hydrolysed ions of uranium are formed. The precipitation of $UO_2(OH)_2$ starts at different pH values depending upon the concentration of uranium. In the author's investigation, auxilary ligands like EDTA and fluoride are added which form weak complexes with uranium which can prevent the hydrolosis. So, under the experimental conditions, the reacting species with dibenzoylmethane has been considered as $Uo_2.2x$. where X is any auxilary ligand. The extraction of uranium (VI) into varying concentrations of dibenzoylmethane into 15% pyridine + 10% n-butanol in chloroform were studied and log-log plots indicated a ratio of 1:2 for uranium to dibenzoylmethane in the extracting species. The extraction process can be expressed as :

$UO_2.2X + DBM \rightleftharpoons UO_2(DBM)_{org}$ ——— (10)

$UO_2(DBM) + (R)$ neutral donor like pyridine or alcohol $\rightarrow [UO_2(DBM)_2(R)]_n$. ——— (11)

So, the D of extraction Process of U(VI)-DBM with Pyridine and n-butanol (neutral donors) can be expressed by the following equation :

$$D = \frac{UO_2(DBM)_2org}{UO_2^{2+}(aq)+UO_2(DBM)^+aq+UO_2(DBM)_2(aq)} \quad \text{——} \quad (12)$$

$$D = \frac{K_d\,\beta_2[DBM]^2}{1+\beta_1[DBM]^2+\beta_2[DBM]} \quad \text{——} \quad (13)$$

References

1. J. Rydberg and B. Bernstrom, *Acta Chem. Scand*, 11, 86; 1957.

2. R.K. Warner, *Australian J. Appl. Sc.*, 4, 427; 1953.

3. A Cacciari, R. De Leone, C. Fizzotti and M. Gabalio, *Energia Nucleare*, 3, 176; 1956.

4. V. Vesely, H. Beranova and J. Maly, *Collection Czech. Chem. Commun.*, 25, 2622;

1960.

5. F.L. Culler, *Ist Int. Conf. Peaceful uses of Atomic Energy, Geneva*, 1955, S. 822

6. F.R. Bruce, *Ibid*, S. 719

7. F.R. Bruce, J.M. Fletcher, H.H. Hyman and J.J. Katz Progress in Nuclear Energy. Ser-III, *Process Chemistry, McGraw Hill*, New York, 1956.

8. W.H. Hardwick and M. Moreton-Smith, *Analyst*, 83, 9, 1958.

9. P.N. Paley, *Ist Int. Conf. Peaceful uses Atomic Energy*, Geneva, Paper 629; 1955.

10. O.A. Nietzel and M.A. Desesa, *Anal. Chem.*, 26, 756; 1957.

11. D.F. Peppard, G.W. Mason and M.V. Gergel, *J. Inor. Nucl. Chem.*, 3, 370; 1957.

12. T. Ishimori, K. Watanabe and E. Nakamura, *Bull Chem. Soc.*, Japan, 33, 636; 1960.

13. T. Ishii and T. Takeuchi, *Bunseki Kagaku*, 195, 10, i, 125.

14. Norio Ichinose, *Talanta* 18, 21; 1971.

15. J. Korkisch and H. Hubner - *Talanta* 23, 283; 1976.

16. E. Peligot, *Ann. Chim. Phys.* 5:1; 1842.

17. R. Bock and E. Bock, Z Anor. *Allg. Chem.*, 263:146; 1950, (46:7453).

18. A.G. Kurnakova and A.V. Nikoalev, *Zh. Neorg. Khim*, 1958, 3:1028. (52:17930).

19. A Norstrom and L.G. Sillen, *Svensk Kem.* Tidskr. 60:227; 1948. (43:3267).

20. V.M. Vdovenko and E.A. Smirnova, *Radio Khimiya*, 1:521; 1959. (54:13817)

21. H.P. Holcomb and J.H. Yoe, *Anal. Chem.*, 32:612; 1960. (54:19304).

22. L.I. Katzin and J.C. Sullivan, *J. Phys. Colloid Chem.*, 55:336; 1951. (45:5013).

23. V.M. Vdovenko, A.I. Gasanov, L.G. Mashirov, I.G. Suglobova and D.N. Suglobov, *Dokl. Akad. Nauk*, SSSR, 182:585; 1968. (70:14901).

24. T. Ishimari et. al. "Data of Inorganic Solvent Extraction" Res. Rept. *Japan At. Energy Res. Inst.*, 1963. JAERI; 1047, 1964; JAERI 1062; 1966. JAERI 1106. (62:13922; 64:15070; 65:6638).

25. G.F. Best, E. Hesford, and H.A.C. McKay, *Congr. Int. Chim. Pure Appl.*, 16e Paris, 1957, Mem. Sect. Chim. Minerale, 1958, 389, (55:15088).

26. M. Branica and E. Bona, Proc. U.N. *Int. Conf. Peaceful uses Atomic Energy*, 2nd Geneva, 1958, 17:172. (54:20605).

27. E. Soerensen, *Acta Chem, Scand*, 1960, 14:965. (56:66558).

28. W.J. Maeck, G.L. Booman, M.C. Elliott, and J.E. Rein, *Anal. Chem.*, 1959; 31:1130. (AA7:977).

29. V. Sraier, *Jaderna Energie*, 1965, 11:52. (62:12772).

30. T. Sato, T. Nishida and M. Yamatake, *J. Appl. Chem. Biotechnol.* 1973; 23, 909. (81:69104).

31. J.C. White *U.S. At. Energy Commn.* 1956, ORNL - 2161 (51:2405).

32. C.A. Horton and J.C. White Anal. Chem. 1958; 30, 1779. (55:1996).

33. B.N. Laskorin, D.I. Skorovarov, and V.V. Shatalov, *Ekstraktsiya Teoriya, Primenenie, Appature, Sb. Statei*, 1962; 163 (59:3601).

34. W.D. Shults and L.B. Dunlop, *Anal. Chim. Acta*, 1963, 29:254. (59:12174).

35. B.N. Laskorin, L.A. Fedorova, V.V. Shatalov and N.P. Stupin, *Dakl. Akad. Nauk, SSSR*, 1967, 174:1334. (68:33708).

36. K. Watnabe, Res. Rept. *Japan Atomic Energy Res. Inst.* 1967, JAERI, 1132. (67:68140).

37. B.N. Laskorin, D.I. Skorovarov, L.A. Fedorova and V.V. Shatolov, *At, Energy*, 1970. 28:383. (73:92156).

38. A.V. Nikolaev, V.B. Durasov, and Z.N. Miranova, *Dakl. Akad. Nauk, SSSR*, 1969, 186:603. (71:85123).

39. L.L. Burger, U.S. At. *Energy Comm.* 1957, H.W. - 44888 (51:14585).

40. T.V. Healy, and J. Kennedy, *J. Inor. Nucl. Chem.* 1959; 10:128; T.V. Healy, J. Kennedy, and G.M. Waind, *ibid*, 137. (53:16656).

41. B.N. Laskorin, L.A. Fedorova, N.P. Stupin, M.I. Kabachink. T. Ya. Medved, and Ya. M. Pokilaropov, *Radio Khimiya*, 1970. 12:335 (73:39258).

42. V.G. Torgov, A.V. Nikolaev, V.A. Mikhailov, L.G. Stadnikova and I.L. Kotlyarevskii, *Zh. Neorg. Khim*, 1968. 13:228. (68:81802).

43. A.M. Rozen, Yu. I. Murinov, Yu. E. Nikitin, and V.P. Krivonogov, *Radio Khimiya*, 1971; 13:891. (76:145459).

44. K. Watanabe, *J. Nucl. Sci Technol.* (Tokyo); 1964. 1:155. (62:8620).

45. T. Sato, *J. Appl. Chem.*, 1966. 16:53. (64:15072).

46. H. Menke and G. Herrman, *Z. Anal. Chem.*, 1960. 175:324. (55:3290).

47. T.J. Hayes and A.G. Hamlin, *Analyst*, 1962. 87:770. (58:921).

48. M. Irving and D.N. Edington, *J. Inorg. Nucl. Chem.* 1959. 10"306. (53:17745).

49. B.N. Laskorin and V.A. Kuznetsov, *Ekstraktsiya, Teoriya, Primenenie, Apparatura*, Sb. Statei, 1962. (209. (58:13202).

50. V.M. Vdovenko, A.A. Liporskii, S.A. Nikitina, and N.E. Yakovleva, *Radio Khimiya*, 1965. 7:509. (64:18490).

51. R.P. Larsen and CA. Seils, Jr., *Anal. Chem.*, 1960, 32:1863 (60:8624).

52. G.V. Ellert, G.T. Bolotova, and T.I. Krasovskaya, *Zn. Neorg. Khim.*, 1968, 13:1386). (69:30700).

53. E. Szabo, A Balazs, and L. Bakos, Solvent Extr. Chem., Proc. *Int. Conf. Goteborg*, 1966, 236. (69:70577).

54. A.M. Rozen, Z.I. Nikolotova, N.A.. Kartasheva, A.I. Zarubin, and E.G. Teterin,

Radio Khimiya, 1971, 13:700 (76:50701).

55. V.M. Vdovenko, I.G. Suglobova, and D.N. Suglobov, *Radio Khimiya*, 1968, 10:182. (69:80894).
56. V.I. Kuznetsov, and I.V. Seryakova, *Ekstraktsiya, Teoriya, Primenenie, Apparatura*, Sb. Statie, 1962, 227, (58:8453).
57. R. Bock, *Z. Anal. Chem.*, 1951, 133:110. (45:6956).
58. T.T. Lai and H.T. Lin, *J. Chin. Chem. Soc.* (Taipei), 1961, 8:327, (58:7368).
59. S.I. Sinyakova, and N.S. Klassova, *Zh. Anal. Khim*, 1959, 14:451 (54:11810).
60. R.N. Shchelokov, I.M. Shulgina, and I.I. Chernyoev, *Dokl. Akad. Nauk, SSSR*, 1966, 168:1338 (65:13196).
61. M. Ziegler, H. Winkler, W. Rittner, and H.H. Wilkening, *Naturiwis*, 1967, 54:141 (68:6880).
62. Cl. Toussaint and G. Cattin, *Analusis*, 1972, 1:347(78:37596).
63. H. Onishi and Y. Toita, *Bunseki, Kagaku*, 1969, 18:1134. (72:28138).
64. F. Vogliotti, *Energia Nucl.* (Milan), 1960, 7:169. (57:6609).
65. C. Deptula, *Inst. Nucl. Res.* (Warsaw) 1966, Rep. No. 754. (66:119354).
66. G.V. Ellert, G.T. Bolotova, and T.I. Krasovskaya, *Zh. Neorg. Khim.* 1971, 16:789. (74:131151).
67. T. Ishimoriand E. Nakamura, *Bull. Chem. Soc.*, Japan 1959, 32:713 (54:12732).
68. J.L. Woodhead and H.A.C. Mckay, *J. Inorg. Nucl. Chem.* 1965, 27:2247: (63:14359).
69. V. Ya. Vasilev, N.N. Andrichuk and A.G. Rykov *Radio Khimiya*, 1972, 14:145 (76:145511).
70. J.S. Fritz and D.C. Kennedy, *Talanta*, 1970, 17:837. (73:136993).
71. V.I. Bykhovtsov, *Radiokhimiya*, 1972, 14:165. (76:158913).
72. M. Beran, Solvent Extr Res., proc. *Int. Conf. Solvent Extr.* Chem. 5th 1968, 75. (74:16207).
73. V. Baran, M. Tympl. and E. Cakrt, *collect. Czech. Chem. Commun.*, 1966, 31:2382 (65:1473).
74. Yu. G. Frolov and G.I. Nasonova, Solvent Extr. Proc. *Int. Solvent Extr. Conf.* 1971, 1:50 (76:15904).
75. J.M.P.J. Verstegen, *J. Inorg. Nucl. Chem.* 1965, 27:201 (62:7166).
76. F. Mannone, C. Stoppa, and G. Sanso, *A.E.C. Accession* No. 4241, 1964, Rept. No. EUR - 1913 i. (62:14130).
77. I.L. Jenkins, H.A.C. McKaby, C.G.C. Shorthill, and A.G. Wain, U.K. *At. Energy Authority Res. Group, At. Energy Res. Eslab Report* 1965, AERE - R - 4440, (63:12618).

78. F. Baronceii, G. Scibona, and M. Zifferero, *J Inor. Nucl. Chem.*, 1962, 24:547 (58:3950).

79. A.M. Rozen and Z.I. Nagnibeda, *Radio Khimiya*, 1971, 13:284 (75:91707).

80. V.S. Sokolov, V.S. Shmidt, E.A. Mezhow and V.N. Shesterikov, *Radiokhimiya*, 1971, 13:459. (75:91649).

81. Z.B. Maksimovic and R.G. Puzic, *J. Inorg. Nucl. Chem.* 1972, 34:1031 (76:117987).

82. E.R. Schmid and E. Juenger, *Radio Chim. Acta*, 1972, 17:153. (78:34505).

83. G. Koch. and E. Schwind, *J. Inorg. Nucl. Chem.*, 1966, 28:571 (64:11939).

84. V.P. Zhivopistsev, B.I. Petrov, E.A. Selezneva, and N.F. Sibiryakov, Tr. Kom. *Anal. Khim., Akad Nauk SSSR*, Inst. Geokhim, Anal. Khim. 1965, 16:80.(69:56698).

85. T. Sato, *Anal. Chim. Acta*, 1969, 45:71 (70:81491).

86. V.M. Vdovenko, M.G. Kuzina, and A.A.. Lipovskii, *Radio Khimiya*, 1971, 13:705. (76:37935).

87. M. Ziegler, *Naturwis*, 1959, 46:353. (53:21354).

88. J. Vera Palomino, A. Bellido, Gonzalez, J.A. Galiano Sedano and R. Parallada Bellod, *An. Quim.*, 1970, 66, 901. (74:68373).

89. T. Sato and M.L. Good, Solvent Extr. *Chem. Proc. Int. Conf.*, Goteborg, 1966, 477, (69, 80913).

90. V.M. Tarayan, E.N. Ovsepyan and A.A. Petrosyan, *Zh. Anal. Khim* 1971, 26:322. (74:134449).

91. A.A. Lipovskii and M.G. Kuzina, *Zh. Neorg. Khim*, 1972, 17:1127. (77:10253).

92. A. Sato and S. Suzuki, *Bull. Chem. Soc.*, Jap., 1970, 43:1380. (73:39241).

93. D.H. Liem, *Acta Chem. Scand*, 1968, 22:773. (69:30680).

94. G.W. Mason, N.L. Schofer, and D.F. Peppard, *J Inorg. Nucl. Chem.* 1970, 32:3899. (74:46761).

95. B.N. Laskorin, and V.F. Smirnov, *Zh. Prikl. Khim*, (Leningrad) 1965, 38:2232 (64:2801).

96. G.W. Mason, S. Lewey and D.F. Peppard, *J. Inorg Nucl. Chem.* 1964, 26:2271 (62:7170).

97. K. Kimura, *Nippon Genshinyoku Gakkaishi*, 1960, 2:585 (55:21769).

98. A.M. Kiwan and R.S. Amin, *J. of Inor. Nucl. Chem.* 1973, 35:3315, (79:140161).

99. T. Sato, *J of Inor. Nucl. Chem.* 1963, 25:109 (58:9672).

100. A.M. Rozen, B.V. Martynov, and V.I. Anikin, *Radio Khimiya*, 1973, 15:24 (78:115812).

101. E.R. Schimid and W. Pfannhauser, *Atom Kernenergie*, 1972, 19:5. (76:158979).

102. F.E. Butler, *Anal. Chem.* 1965, 37:340 (62:13504).

103. V.L. Bykhovtsov and T. Ya. Zimina, *Radio Khimiya*, 1970, 12:686. (74:46172).

104. J. Rydberg and B. Rydberg, *Ark. Kemi*, 1955, 9:95 (50:6245).

105. A. Krishen and H. Freiser, *Anal. Chem.*, 1959, 31:923 (5.:13868).

106. D. Dyrssen, *Nucl. Sci. Eng.*, 1963, 16:448, (60:2384).

107. J. Rydberg, *Ark. Kemi*, 1955, 8:101 (49:13006).

108. T. Shigematsu and M. Tabushi, *Bull. Inst. Chem* Res. Kyoto Univ., 1961, 30:35 (56:3172).

109. M. Tabushi, *Bull. Inst. Chem.* Res. Kyoto Univ., 1959, 37:237 (54:743).

110. V.V. Bagreev and Yu. A. Zolotov, *Zh. Anal. Khim* 1965, 20:867 (63:17215)

111. K. Batzar, D.E. Goldberg, and L. Newman, *J. Inor. Nucl. Chem.*, 1967, 28, 1511. (67:47720).

112. D.B. Stevancevic and V.G. Antonijevic, Bull. Boris Kidric, *Inst. of Nucl. Sci*, 1965, 16:11 (63:17215).

113. J. Stary and E. Hladsky, *Anal. Chem., Acta*, 1963, 28:227 (58:13107).

114. J.L. Long and L.F. Grill, *U.S. At. Energy Comm.*, 1957, RFP-78 (54:19304).

115. Y. Umezaki, *Bull. Chem. Soc.*, Jap, 1963, 36:769. (59:9322).

116. J. Stary, *Collect Czech. Chem. Commun.*, 1960, 25:890> (54:13981).

117. T. Shigematsu and M. Tabushi, *Nippon Kogaku*, Zassni., 1960, 81:265. (54:13981).

118. V. Moncka and J. Stary, *Collect Czech. Chem. Commn.* 1961, 26:763. (55:15088).

119. R. Fontaine, B. Santoni, C. Pommier, and G. Guichon, *Anal. Chim., Acta*, 1972, 62:337. (78:37605).

120. B.S. Jensen Solvent Extr. Res. Proc. *Int. Conf. on Solvent Extr. Chem.*, 5th 1968, 29, (74:16203).

121. S.A. Pai, Report 1970, *INIS* - mf-360. (81:17247).

122. R.H. Betts an d R.K. Leigh, *Can. J Res.*, 1950, 288:514. (45:3746).

123. S.M. Khopkar and A.K. De., *Chem. Ind.* (London), 1959, 291, (53:17771).

124. S. Peterson, *J Inor. Nucl. Chem.*, 1960, 14:126 (55:1262).

125. T.V. Healy, *J Inor. Nucl. Chem.*, 1961, 19:314, 528 (56:9502).

126. J.W. Mitchell, *U.S. At. Energy Comm.*, 1970, IS-T-353. (74:116604).

127. K. Katsura and M. Shinagawa, *Nippon Genshiryoku* Gakkaishi, 1972, 14:100 (76:147849).

128. D. Dyressen, Trans. Roy. *Inst. Technol.*, Stockholm, 1962, No. 188, (57:2915).

129. J. Stary, *Anal. Chim. Acta*, 1963, 28:132 (58:9610).

130. K.S. Koppiker and K.B. Gajankush, *Indian At. Energy Comm.*, At Energy Estab., 1965, Trombay, AEET-225, (64:4254).

131. K. Motojima, H. Yoshida, and T. Imahashi, *Bunseki Kagaku*, 1965, 11:1028. (58:19).

132. C.L. Rulfs, A.K. De, J. Lakritz, and P.J. Elving, *Anal. Chem.*, 1955, 27:1802. (50:2365).

133. K. Motojima, H. Yoshida and K. Izawa *Anal. Chem.*, 1960, 32:1083. (54:24125).

134. K. Motojime and H. Hashitani, *Bunseki Kagaku*, 1960, 9:151. (55:17359).

135. D. Dyrssen, M. Dyrssen and E. Johansson, *Acta Chem.*, Scand, 1956, 10:106. (50:13565).

136. I.P. Alimarin, and Yu. A. Zolotov, *Zh. Anal. Khim*, 1957, 12:176. (52:165).

137. G.A. Saed, N.P. Rydenko, and A.V. Lapttsky, *Z. Anorg Allg. Chem.*, 1966, 347:327. (66:32381).

138. N. Saito and T. Sekine, *Bull. Chem. Soc.*, Japan, 1958, 31:789. (53:9840).

139. J. Stary and J. Smizanska, *Anal. Chim. Acta*, 1963, 29:545. (60:2384).

140. S.J. Kim, C.J. Yoon, and I.S. Chang, *Dachan Hwahak Hwoejee*, 1969, 13:16 (72:16159).

141. N.P. Rudenko, A.G. Saed, and A.V. Lapitskii, *Radio Khimiya*, 1965, 7:142 (63:12394).

142. J.P. Shukla, Y.K. Agrawal, and K. Bhatt, *Separ. Sci.*, 1973, 8:387. (79:13217).

143. Y.K. Agrawal, *Separ. Sci.*, 1973, 8:709. (80:33554).

144. F. Baroncelli and G. Grossi, *Com. Naz. Energy Nucl.* 1970, RT/CHI (70) 26 (74:116613).

145. S. Shibata, *Anal. Chim. Acta*, 1960, 22:479. (54:13981).

146. K.L. Cheng, *Talanta*, 1962, 9:739. (57:10527).

147. K.R. Solanke and S.M. Khopkar, *Chem. Anal.* (Warsaw), 1972, 17:1175. (78:118821).

148. R.A. Zingaro, *J Amer. Chem. Soc.*, 1956, 78:3568. (50:16527).

149. B.V. Agarwal, S.P. Sungal, and A.K. Dey, *Z. Anal. Chem.*, 1965, 207:256. (62:8385).

150. H. Bode and F. Neumann, *Z. Anal. Chem.*, 1960, 172:1 (54:8467).

151. A. Traub and D.F. Boltz, *Mikrochim. Acta*, 1969, 749: (71:67121).

152. P. Pakalns and B.R. Mc. Allister, *Anal. Chim., Acta*, 1972, 62:207. (78:52242).

153. Seaborg, G.T. and J. Katz., ' *The Actinide Elements* - New York, McGraw Hill, 1954.

154. Ahrland S. *Acta Chim. Scand*, 3, 374, 1949.

3

Extraction - Spectrophotometric Studies on the U(VI) - DBM-3,5,diBr PADAP Mixed Ligand System

3.1 Introduction

The most commonly used methods for the spectrophotometric determination of uranium are based on reactions with hydrogen peroxide[1], arsenazo-I[2,3], arsenzo-III[4,5], and 8-quinolinol[6]. More recently, procedures based on the formation of ternary uranium complexes with Pyrogallol Red and cationic surfactants[7,8] have been suggested. Chromal blue G (Sodium-2"-chloro-4"-nitro-4-hydroxy-3,3'-dimethyl fuchsone - 5.5' - dicarboxylate; C.I. - 43835), a triphenyl methne dye, has been used as a spectrophotometric reagent for beryllium[9], Scandium[10], Palladium[11] and gallium[12]. It has also been found that microgram amounts of uranium react very sensitively with chromal blue G in the presence of a cationic surfactant[13].

Spectrophotometric and fluorimetric methods[14] are employed extensively for the determination of uranium in a variety of geological samples, including materials of very low uranium contents such as basaltic rocks and of uranium rich materials, as for instance torbernites, carnotite, uraninite and pitchblende. For the successful analysis of both types of uranium-bearing minerals the very sensitive spectrophotometric method of uranium determination can be used, which is based on measurement of the absorbance of the green uranium(IV)-arsenazo-III complex in moderately concentrated hydrochloric acid medium. Interferences are caused by thorium, zirconium, titanium, rare earth elements molybdenum and iron. So it is necessary to separate uranium from these elements and also from other components such as the main constituents of minerals and rocks. For this purpose, methods based on liquid - liquid extraction and ion-exchange have found widespread application. Pyrogallol Red (PGR)[15,16] forms a violet chelate with the uranyl ion. Other metallochromic indicators widely used for the spectrophotometric determination of uranium(VI) include PAR[17] PAN[18], Eriochrome cyanine R[16] and chromazurol S.[19,20]. These colour reactions are not specific and the sensitivity is also not adequate especially for very small quantities of uranium(VI).

In the two-cycles solvent extraction di(2-ethylhexyl), phosphoric acid-n-trioctyl phosphine oxide (DEHP-TOPO)_ Process[21,22], the lack of methods for determination of uranium in real system, i.e., in aqueous ammonium carbonate solution exist. Despite the abundance of methods for the uranium analysis, very few applications[23] of these methods have been reported for the determination of uranium in carbonate solution.

The exploitation of low-grade ores involves processes in which the control of uranium at trace levels in leach liquors and effluents is important and methods on-and off-line, are required for the determination of uranium at microgram to nanogram levels.

Pyridylzo compounds with a hydroxy group in the ortho position with respect to the azo group are valuable chromogenic reagents[24]. They have been used widely in compleximetric and extraction photometric methods for the determination of many ions. Highly sensitive regents have been obtained by the introduction of the diethyl-amino group, which is a strongly electron-donating group in the para position next to the azo group. The azo dye, 2-(2-Pyridylazo)-5-diethylamino phenol (PADAP) and its bromine substituted derivative 2-(5-bromo-2-pyridylazo)-5-diethylaminophenol (Br-PADAP), the latter having a greater tolerance to iron(III), have found wide application as very sensitive reagents in the spectrophotometric determination of uranium (VI)[24-34]. Johnson and Florence[28] found that in the presence of fluoride a stable 1:1:1 ternary fluoride complex {Uo_2(Br-PADAP)F} was formed at pH values between 6 and 8. At molar ratios of fluoride to uranium above 2000:1, the complex formation became independent of the fluoride concentration. The molar absorptivity of the fluoride complex at 578 nm was 7.32×10^4 l mol $^{-1}$ cm $^{-1}$, whereas that of Uo_2 (Br-PADAP) was 5.63×10^4 l mol^{-1} cm^{-1}.

It is observed in the author's investigations that the selectively extracted colourless U(VI)-DBM complex can be converted into an intensely coloured ternary complex by reaction with chromogenic complexing agents like 4-(2-Pyridylazo) resorcinol (PAR), 2-(2-pyridylazo)-5-diethylmino Phenol (PADAP), 2-(5-bromo-2Pyridylazo)-5-diethylminophenol (Br-PADAP), 2 - (3.5-dibromo-2-pyridylazo)-5-diethylamino Phenol 3,5-diBr(PADAP) and 1-(2-Pyridylazo) - 2 -naphthol (PAN) in presence of a surfactant medium like Pyridine + alcohol which can offer micellar phase. These chromogenic complexing agents though can react with uranium(VI) in aqueous phase directly to produce coloured complexes their analytical applications are severely limited as many metal ions like iron, Pb, Sn, Ag, Cu and Ni interfere seriously and require elaborate separation procedures. Further the colour reaction are less sensitive in aqueous phase.

In view of the importance of the determination of uranium(VI), at trace levels when associated with different interfering metal ions the author has taken up a systematic study to develop a highly selective separation of uranium(VI) by extraction with DBM into 15% pyridine + 10% n-butanol in chloroform mixed solvent from matrix components in presence of masking agent mixture (EDTA + Fluoride) followed by ternary complex formation in chloroform + n-butanol + pyridine medium, by reacting an aliquot of the extract by addition of 2-(3.5-dibromo-2-Pyridylazo)-5-diethylamino Phenol solution in ethanol medium which produces intensely coloured species. This method is free from many interferences from variety of diverse ions. To exploit this reaction system for the precise trace determination of uranium when present in various diversified complex matrix components, conditions are optimised based on detailed and systematic investigations. The corresponding results are described in the following pages.

EXPERIMENTAL

3.2 Reagents

(a) PAR Solution : 0.2% (m/v) : 237.2 mg. of the sodium salt of PAR is dissolved in 100 ml of absolute ethanol.

(b) PADAP and Bromo-PADAP Solution : - >0.005% m/v. 0.05 g. of bromo-PADAP is dissolved in 100 ml. of absolute ethanol. Fresh solution are prepared every week.

(c) 3,5 di Bromo-PADAP Solution 0.02gm of 3,5-dibromo - PADAP (Fluka) is dissolved in 100 ml. of absolute ethanol.

All other solutions are prepared as described in the previous chapter.

3.3 General Procedure :

To 5.0 ml of 20 μg/ml of uranium(VI) Solution, transferred into a 125 ml separatory funnel, add 0.5 ml. each of the following masking agent solutions (15% EDTA and 2% sodium fluoride) and allow to stand for 2 minutes. To this add, 2.5 ml each of 1% calcium nitrate and 5% aluminium nitrate salting out agent solutions. Make up the total volume to 25 ml with distilled water, maintaining the pH at 6.5 by adding dilute sodium hydroxide or hydrochloric acid. To this add 5 ml of 0.2% dibenzoylmethane in chloroform and 20 ml of (15% Pyridine + 10% n-butanol in chloroform) mixed solvent and equilibrate for 2 minutes.

After clear separation of the two phases, quantitatively remove the organic extract and transfer an aliquot (2.0 ml) of this to a 25.0 ml volumetric flask, add 2.0 ml of 0.05% 2-(3.5-dibromo-2-Pyridylazo)-5-diethylamino Phenol 3,5 diBr PADAP in ethanol and heat on water bath for 30 minutes. Cool it to the room temperature and make upto the mark with dimethyl formamide (DMF). Measure the absorbance at 590 nm against a similarly prepared reagent blank.

3.4 Selection of Best Chromogenic Complexing Reagent for Ternary Complex Formation

It is observed that U(VI)- DBM complex extracted into 15% Pyridine + 10% n-butanol in chloroform mixed solvent can be converted into a ternary complex of high colour sensitivity by adding a chromogenic complexing agent like 4-(2-pyridylazo) resorcinol (PAR), 2-(2-pyridylazo)-5-diethylaminophenol (PADAP), 2-(5-bromo-2-Pyridylazo)-5-diethylaminophenol (Br-PADAP), 2-(3.5-dibromo-2-Pyridylazo)-5-diethylaminophenol (3,5 di Br PADAP). Of these reagents tested best results with good reproducibility and high sensitivity are observed with 2-(3.5-dibromo-2-Pyridylazo)-5-diethylamino Phenol (3,5 diBr PADAP) which is used in all further investigations. The corresponding results are given in Table 3.1

3.5 Effects of Composition of Solvent Medium

The effect of composition of solvent medium on the colour intensity of (U(VI)-DBM-3,5diBr(PADAP) ternary complex is examined by studying the colour reaction in different organic media by varying the solvent medium for making up the coloured complex in 25.0 ml flask, after heating it on hot water bath for 30 minutes. The results of the investigations are given in Table 3.2. It is observed that when 3.5 diBr (PADAP) in ethanol is used in the DMF as a diluent maximum absorbance is obtained for the U(VI)-DBM-3,5diBr (PADAP) system.

TABLE 3.1

Effect of Various Complexing Chromogenic Reagents on the Absorbance of U(VI)-DBM Complex Extracted into 15% Pyridine + 10% n-Butanol in Chloroform Mixed Solvent

Aqueous Phase : 5.0 ml. of U(VI) (20 μg/ml) + 0.5 ml each of 15% EDTA and 2% sodium fluoride masking agent solution + 2.5 ml each of 1% calcium nitrate and 5% aluminium nitrate salting-out agent solution. Total Volume 25 ml. pH: 6.

Organic Phase : 5.0 ml. of 0.2% dibenzoylmethane in chloroform + 15% pyridine + 10% n- butanol in chloroform, total volume 25 ml.

Non-aqueous Medium : 2 ml of organic extract + 2 ml of 0.02% chromogenic agent, heating for 30 minutes, made up to 25 ml with DMF.

S. No.	*Chromogenic Reagent*	*Max (nm)*	*Molar absorptivity liters mole^{-1} cm^{-1}*
1.	4(2-Pyridylazo) resorcinol	530	2.10 x 10^4
2.	2-(2-Pyridylazo)-5-diethylamino Phenol	560	5.0 x 10^4
3.	2-(5-bromo-2-Pyridylazo)-5-diethylamino Phenol	574	6.33 x 10^4
4.	2-(3.5-dibromo-2-Pyridylazo) -5-diethylamino Phenol	590	9.6 x 10^4
5.	1-(3-Pyridylzo)-2-naphthol	570	2.3 x 10^4
6.	Dibenzoylmethane only	400	1.78 x 10^4

3.6 Effect of Concentration of 3,5-diBr PADAP

The effect of variation of concentration of 3,5 diBr (PADAP) on the absorbance of U(VI)-DBM-3.5 diBr PADAP mixed ligand complex in ethyl alcohol medium is investigated using different quantities of 3,5 diBr (PADAP) adopting the general procedure. The results obtained are given below in Table 3.3

It can be seen from the data that 2.0 ml x 10^{-4} M 3,5 diBr (PADAP) in ethyl alcohol is sufficient for attaining maximum absorbance and thereafter addition upto 5 ml shows decrease in absorbance, when measured against the appropriate reagent blank.

Table 3.2
Effects of Composition of Solvent Medium

Aqueous Phase : 5.0 ml. of U(VI) (20 μg/ml) + 0.5 ml each of 15% EDTA and 2% sodium fluoride masking agent solution + 2.5 ml each of 1% calcium nitrate and 5% aluminium nitrate salting-out agent solution. Total Volume 25 ml. pH: 6.

Organic Phase : 5.0 ml. of 0.2% dibenzoylmethane in chloroform + 15% pyridine + 10% n- butanol in chloroform, total volume 25 ml.

Non-aqueous Medium : 2.0 ml. of organic extract + 2 ml of 0.05% 3,5 di Br (PADAP) in various non-aqueous media, for 30 minutes, total volume 25 ml with DMF.

S. No.	*Chromogenic agent 0.5%*	*Absorbance at 590 nm make up with ethylalcohol*
1.	3.5 di Br (PADAP) in dimethyl formamide	0.65
2.	3.5 di Br (PADAP) in Ethyl alcohol	0.61
3.	3.5 di Br (PADAP) in Dichloroethane	0.31
4.	3.5 di Br (PADAP) in Ethyl Acetate	0.45
5.	3.5 di Br (PADAP) in Acetone	0.23
6.	3.5 di Br (PADAP) in Isomylalcohol	0.44

3.7 Absorption Spectra of U(VI) - DBM-3,5 - diBr PADAP.

The absorption spectra of U (VI) - DBM-3,5 diBr PADAP mixed ligand complex and DBM-3,5 diBr PADAP reagent blank in chloroform ethyl alcohol-DMF medium are recorded as described below :

5.0 ml of 8.4 x 10-5 M uranium (VI), 0.5 ml each of 15% EDTA and 2% sodium fluoride are placed into a 125 ml. separatory funnel and allowed to stand for 2 minutes. To this 2.5 ml each of 1% calcium nitrate and 5% aluminium nitrate solution are added and made upto 25.0 ml maintaining the pH at 6.5 by adding dilute sodium hydroxide or hydrochloric acid. To this 5 ml of 0.2% dibenzoylmethane in chloroform and 20 ml of 15% pyridine and 10% n-butanol in chloroform are added

TABLE 3.3

Effect of Concentration of 3,5 diBr-PADAP in Ethyl Alcohol on the Absorbance of U(VI)-DBM-Pyridine-3,5 Di Br-PADAP Mixed Ligand Complex

Aqueous Phase : 5.0 ml of U(VI) (20 μg/ml) + 0.5 ml each of 15% EDTA and 2% sodium fluoride masking agent solution + 2.5 ml each of 1% calcium nitrate and 5% aluminium nitrate salting out agent solution, total volume 25 ml, pH:6.

Organic Phase : 5.0 ml of 0.2% dibenzoylmethane in chloroform + 15% Pyridine and 10% n-butanol in chloroform, total volume 25 ml.

Non-aqueous Phase : 2.0 ml of organic extract + X.0 ml. of 1 x 10-4 M - 3,5 diBr PADAP in ethyl alcohol. Heat for 30 minutes on hot water bath, make upto 25.0 ml with DMF.

S. No.	*[3,5-diBr PADAP] [U(V)]*	*Absorbance at 590 nm*
1.	0.740	0.27
2.	1.480	0.44
3.	2.98	0.570
4.	4.46	0.65
5.	5.95	0.65
6.	8.93	0.65
7.	11.90	0.65
8.	14.88	0.65
9.	17.86	0.65
10.	20.83	0.65
11.	23.81	0.65

and equilibrated for 2 minutes. After clear separation of the phases, the organic phase is removed and 2.0 ml of this is transferred into 25 ml volumetric flask. To this 2.0 ml of 1 x 10^{-4} M 3,5-diBr PADAP in ethyl alcohol is added. The contents were heated on hot water bath for 30 minutes and allowed to cool, and then made upto the mark with DMF. Then the absorbance is measured at different wavelengths against similarly prepared reagent blank. The absorption spectra of DBM-3,5 diBr PADAP reagent blank is also recorded at different wavelengths against 15% Pyridine + 10% n-butanol in chloroform medium.

It can be seen from Fig. 3.1 that U(VI)-DBM-3,5 diBr PADAP mixed ligand complex shows maximum absorbance at 590 nm and the reagent blank has no significant absorbance at 590 nm. It can be further observed that U(VI)-DBM exhibits maximum absorbance at 400 nm when measured against its reagent blank, as can be seen from Fig. 3.2.

3.8 Beer's Law Applicability

Spectrophotometry is one of the most potential fields in analytical chemistry. In recent times, spectrophotometry received much importance in the assay of trace metals due to its simplicity and reproducibility. Very extensive range of concentrations from macro quantities to traces may be analysed in spectrophotometric methods. These methods are based on the simple relationship between the absorption of monochromatic radiation by a coloured solution and the concentration of the coloured species in the solution.

The fundamental law governing the absorption of all types of electromagnetic radiation is known as the Lambert - Beer's law. The basic law of Spectrophotometry is given by the equation.

$$\log \frac{I_o}{I} = \epsilon bc = A,$$

where I_o = the radiant power of the beam striking the sample.

I = the radiant power of the beam transmitted by the sample.

b = the thickness of the absorbing sample in cms.

c = the concentration of absorbing solution

ϵ = the extinction coefficient

$\frac{I}{I_o}$ = transmittance = T

A = $\log \frac{I_o}{I} = \log \frac{1}{T}$ = Absorbance

The equation shows that the absorbance of a solution is directly proportional to the concentration of the absorbing species, when the length of light path is fixed and directly proportional to the light path when the concentration is fixed.

It is evident that a plot of absorbance at constant 'b' against concentration gives a straight line, if Beer's law is followed, this line passes through the origin since any absorbance due to the solvent is cancelled out in the usual method of making the measurement against

I U(VI) - 3,5 - diBr PADAP λ max : 576
against reagent blank

II U(VI) - DBM - 3,5 di Br PADAP
against reagent blank λ max : 590

III Reagent blank against water
Concentration of Uranium 1.34 X 10^{-6} M

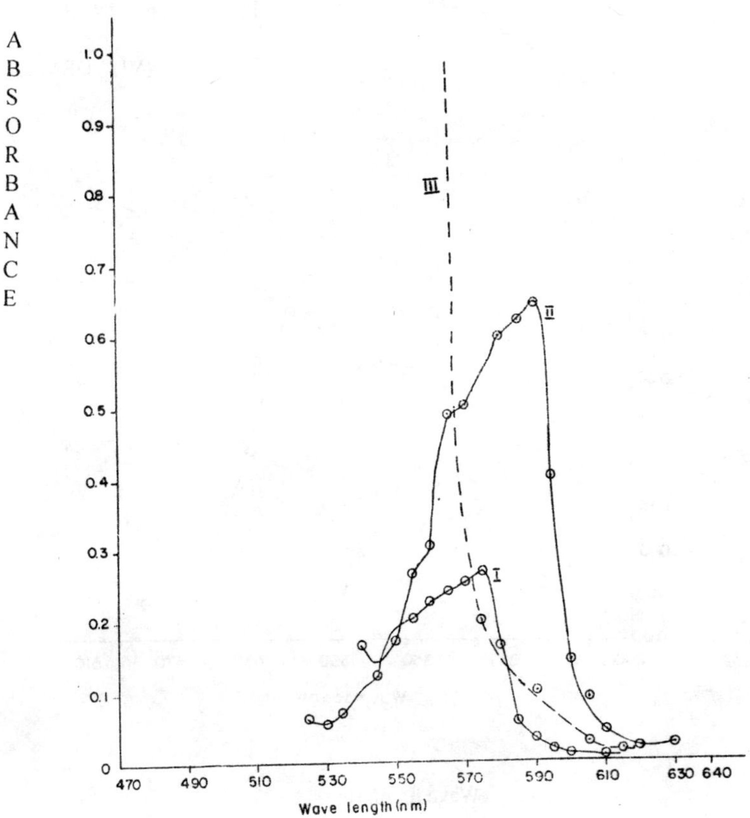

Fig 3.1 : Absorption spectra of Uranium (VI)

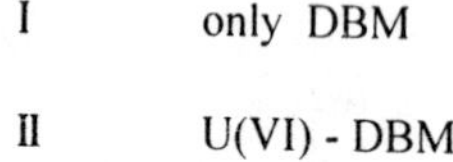

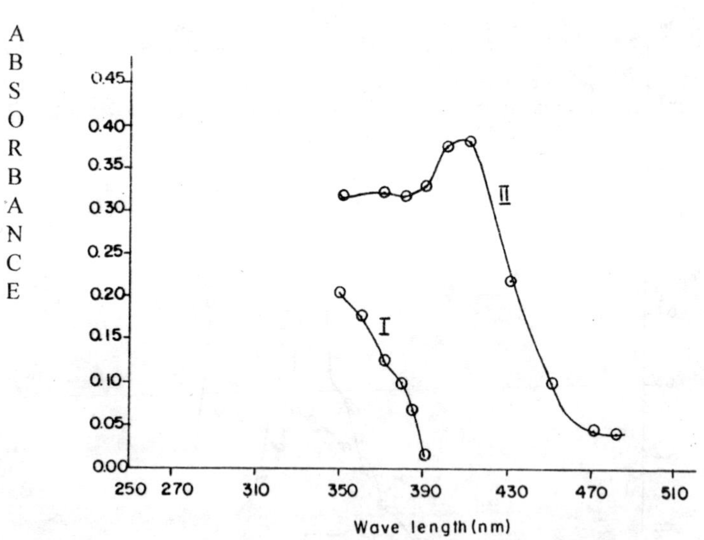

Wave lingth (nm)

Fig 3.2 : Absorption spectrum of U(VI) - DBM Complex

blank, Beer's law is successful in describing the absorption behaviour of dilute solutions only. In this sense it is a limiting law. At high concentrations, the average distance between solute molecules (or ions) is diminished to the point, where each effects the charge distribution of the neighbours. Because the degree of interaction is dependent upon the concentration, the accuracy of this phenomenon causes deviations from the linear relationship between absorbance and concentration.

A second fundamental limitation arises from alterations in the refractive index of the solution as a result of high solute concentrations. Apparent chemical deviations of Beer's law are frequently encountered as consequence of association, dissociation or reaction of the absorbing species with the solvent. Beer's law is obeyed only when the monochromatic radiation is employed which is another limitation to the law

3.9 Adherence of U(VI)-DBM-3,5-diBr PADAP System to Beer's Law

The concentration range for the adherence of U(VI)-DBM-3,5 diBr PADAP system in 15% Pyridine + 10% n-butanol in chloroform medium to Beer's law has been examined by the following procedure :

Procedure

Pipette out an aliquot of uranium (VI) (0-100 μg) solution in a separatory funnel, add 0.5 ml each of 15% EDTA and 2% sodium fluoride solutions and allow to stand for 2 minutes. To this, add 2.5 ml each of 1% calcium nitrate and 5% aluminium nitrate solutions and make up the volume to 25.0 ml with distilled water, maintaining the pH at 6. To this add 5 ml of 0.2% DBM in chloroform and 20 ml of 15% Pyridine + 10% n-butanol in chloroform mixed solvent and equilibrate for 2 minutes. After clear separation of the two phases, transfer the organic layer into a microburette and run 2.0 ml of this solution into a 25 ml volumetric flask. Add 2.0 ml of 1 x 10-4 M 3,5-diBr PADAP solution in ethyl alcohol, and heat on a water bath for 30 minutes. Cool it to the room temperature and make up to the mark with DMF. Measure the absorbance at 590 nm against a similarly prepared reagent blank.

The results obtained are given in Table 3.4 and graphically represented in Fig 3.3. It can be seen that Beer's law is obeyed by U(VI)-3,5 diBr PADAP mixed ligand system upto 7.2 μg/ml uranium (VI).

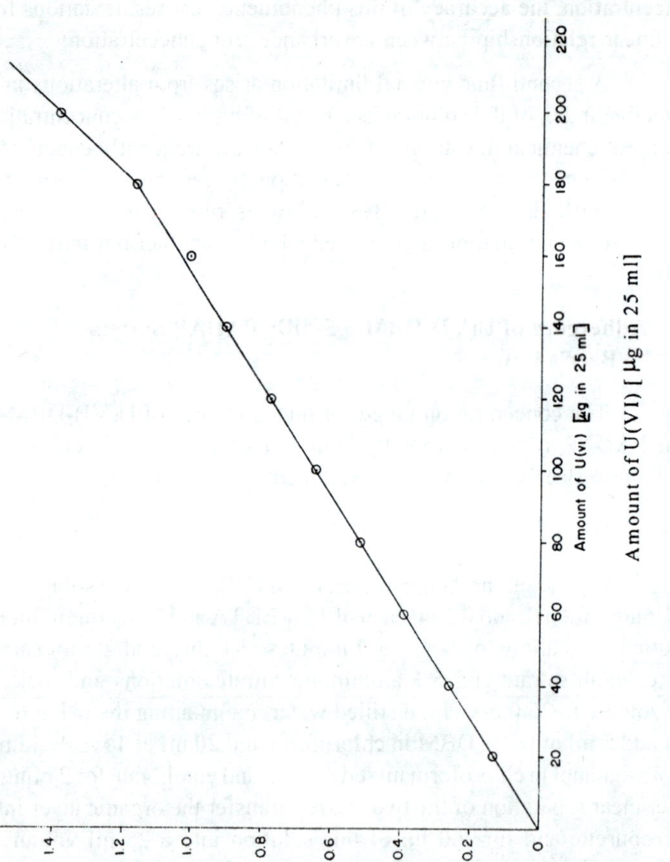

Fig 3.3 : Adherence of U(VI) - DBM - 3,5 - di Br PADAP Methods to Beer's law

TABLE 3.4
Adherence of U(VI)-3,5 diBr PADAP Method to Beer's Law

Aqueous Phase : X.0 ml of U(VI) + 0.5 each of 15% EDTA and 2% sodium fluoride masking agent solution + 2.5 ml each of 1% calcium nitrate and 5% aluminium nitrate salting out agent solution, total volume 25 ml, pH : 6.

Organic Phase : 5.0 ml of 0.2% dibenzoylmethane in chloroform + 15% pyridine and 10% n-butanol in chloroform, total volume 25 ml.

Non-aqueous Phase : 2.0 ml of organic extract + 2 ml. of 1 x 10-4 M - 3,5 diBr PADAP in ethyl alcohol. Heat for 30 minutes on hot water bath, cool. total volume 25.0 ml with DMF.

Sl. No.	*Total amount of uranium present (in the initial aqueous phase, 25.0 ml.) before extraction μg*	*Amount of uranium μg/ml (present in the non-aqueous medium 25.0 ml. used for colour development*	*Absorbance at nm 590*
1.	20	1.6	0.129
2.	40	3.2	0.258
3.	60	4.8	0.387
4.	80	6.4	0.51
5.	100	8.0	0.45
6.	120	9.6	0.774
7.	140	11.2	0.903
8.	160	12.8	1.032
9.	180	14.4	1.161
10.	200	1.60	1.290

3.10 Sensitivity

The Objective numerical expression of the sensitivity of spectrophotometric methods is the molar absorptivity (E) at the wave length of maximum absorbance (λ max) of coloured species.

$\in = A/bc$

If b is in cm, the molar absorptivity ($\in$) is expressed in $l.mol^{-1} cm^{-1}$, if SI units are used. However, if b is in nm, is in $l.mol^{-1}mm^{-1}$ and has 1/10 the value of that expressed in $l.mol^{-1}\ cm^{-1}$. It is convenient to express and compare the sensitivities of spectrophotometric methods in terms of specific absorptivity[35] (a). This is obtained by dividing $\in$ by the atomic weight of the element by 1000.

$$a = \frac{\epsilon}{\text{at.wt. x 1000}}$$

The value a (in ml. g^{-1} cm^{-1}) corresponds to the absorbance of a 1 μg/ml (1 ppm) solution of the determined in cuvette with an optical path length of one cm.

The sensitivity of spectrophotometric method is often expressed in terms of sensitivity index (s) given by Sandell[1] which represents the number of micrograms determined per ml. of a solution having an absorbance of 0.001 for a path length of 1 cm. The sensitivity (s) according to sandell is expressed in $\mu g/cm^2$.

The average molar extinction coefficient and specific absorptivity for U(VI)-DBM-3,5 diBr PADAP mixed ligand complex into 15% pyridine + 10% n-butanol in chloroform mixed solvent medium are found to be $E = 9.1 \times 10^4$ and $a = 0.3824$ ml g^{-1} cm^{-1}. The Sandell's sensitivity of the colour reaction is found to be $0.00262 \mu g/cm^2$.

3.11 Optimum Range from Ayre's[36] and Ringbom's[37] Plot

From the equation $\frac{dc}{c} = \frac{-dI}{I} \times \frac{0.434}{E}$ in which

$$\epsilon = \log \frac{I_o}{I},$$

the percent relative error (dc/c x 100) per one percent photometric error has been calculated and plotted against transmittance (Ayre's plot). The way in which analysis error varies with the transmittance is shown graphically in Fig. 3.4 and Table 3.5

In order to obtain the suitable concentration range, the percent absorbance as ordinate has been plotted against logorithm of concentration as abscissa in Fig 3.5 as proposed by Ringbom's and results are given in Table 3.6.

The curve passed through an inflexion shown in Figure 3.5, the accuracy is greatest where the curve has its steepest slope. i.e.,

$$di/dc/c = \frac{dI}{2.303} \, d \log c = \lambda \max$$

It is evident from the graph that the optimum range with minimum photometric error is in the region 1.6-4.0 ug/ml of uranium(VI) with layer depth of one cm.

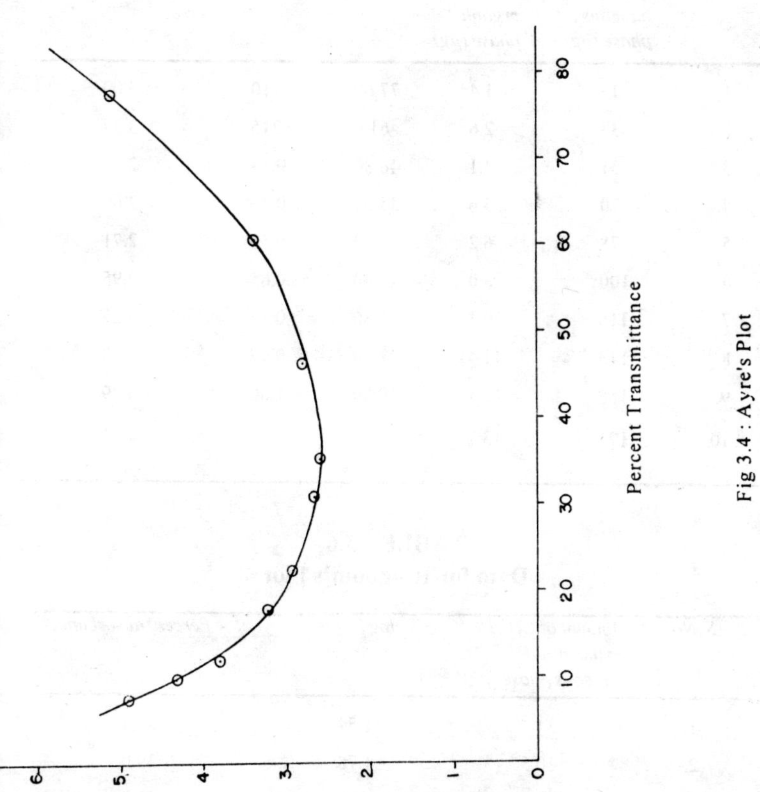

Fig 3.4 : Ayre's Plot

TABLE 3.5
Data for Ayre's Plot

S. No.	*Amount of U(VI)*		*Percent transmit-tance*	*Absorbance*	*dc/c x 100 percent relative error one percent photometric error*
	Initial aqueous phase (μg)	*Final organic phase (μg)*			
1.	17	1.4	77.60	0.110	5.09
2.	33	2.6	61.0	0.215	3.37
3.	51	4.1	46.80	0.33	2.78
4.	70	5.6	35.50	0.45	2.69
5.	78	6.2	31.60	0.50	2.71
6.	100	8.0	22.40	0.65	2.95
7.	116	9.3	17.80	0.75	3.22
8.	143	11.4	12.00	0.92	3.78
9.	155	12.4	10.00	1.00	4.29
10.	171	13.7	7.90	1.10	4.91

TABLE - 3.6
Data for Ringbom's Plot

S. No.	*Amount of U(VI) in the initial aqueous phase (μg)*	*log U(VI)*	*Percent absorbance*
1.	34.7	1.54	22.4
2.	60.5	1.78	39.0
3.	82.5	1.92	53.2
4.	100.0	2.00	64.5
5.	106.0	2.03	68.4
6.	120.3	2.08	77.6
7.	127.4	2.10	82.2
8.	135.5	2.13	87.4
9.	139.5	2.14	90.0
10.	142.8	2.15	92.1

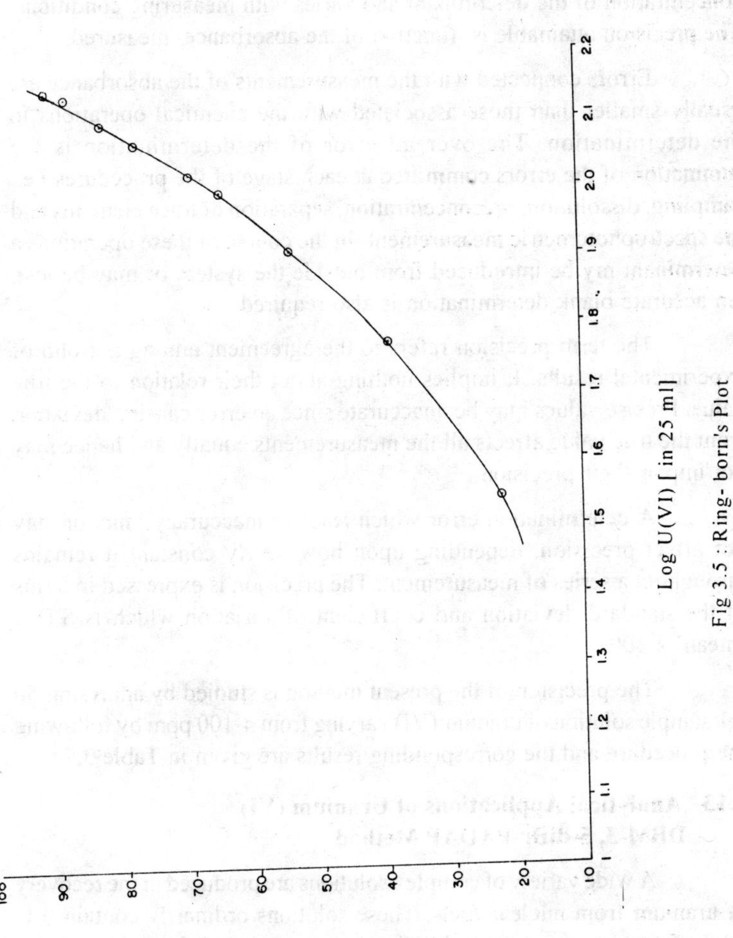

Fig 3.5 : Ring- born's Plot

3.12 Precision

The precision of spectrophotometric methods depends on the concentration of the determinant and varies with measuring conditions. The precision attainable is function of the absorbance measured.

Errors connected with the measurements of the absorbance are usually smaller than those associated with the chemical operations in the determination. The over all error of the determination is the summation of the errors committed at each stage of the procedures i.e., sampling, dissolution, preconcentration, separation of trace elements and the spectrophotometric measurement. In the course of these operations a determinant my be introduced from outside the system or may be lost. An accurate blank determination is also required.

The term precision refers to the agreement among a group of experimental results. It implies nothing about their relation to the true value. Precise values may be inaccurate since an error causing deviation from the true value affects all the measurements equally and hence may not impair their precision.

A determination error which leads to inaccuracy may or may not affect precision, depending upon how nearly constant it remains throughout a series of measurement. The precision is expressed in terms of the standard deviation and coefficient of variation which is S.D. / 'mean' x 100.

The precision of the present method is studied by analysing 50 ml. sample solution of uranium(VI) varying from 4-100 ppm by following the procedure and the corresponding results are given in Table 3.7.

3.13 Analytical Applications of Uranium (VI) DBM-3, 5-diBr-PADAP Method

A wide variety of complex solutions are produced in the recovery of uranium from nuclear fuels. These solutions ordinarily contain 0.1-20 g of uranium per liter and high concentrations of aluminium nitrate in addition to a variety of cations and anions associated with fuel-element constituents and dissolutions media.

Although, accurate determinations of uranium can be made with combinations of such techniques as cupferron precipitation, sulfide precipitation, hydroxide precipitation, solvent extraction, and electrolytic separations of impurities, such procedures must be tailored to the sample and require careful sample preparation and demand great skill from the analyst. In view of potential and serious pollution hazards of soluble U(VI) occuring from various mines, effluents, nuclear rector wastes, it is necessary to monitor U(VI) levels in various surface waters, ground waters and waste waters.

TABLE 3.7

Precision Studies of UVI-DBM-3,5 diBr PADAP Methods-Data of Replicate Analysis of U(VI)

S. No.	Conc. of U (VI)	Replicate (Ug) 1	2	3	4	5	6	Mean (μg)	S.D.	C.V.	Found Mean ±ts / N
1.	4	3.9	3.7	4.2	4.5	4.2	4.4	4.2	0.31	7.38	4.2 ± 0.32
2.	8	8.2	7.8	8.4	7.8	8.4	7.5	8.0	0.37	4.62	8.0 ± 0.39
3.	12	11.6	11.8	12.3	12.2	12.4	12.5	12.1	0.36	2.97	12.1 ± 0.38
4.	16	15.5	15.8	16.2	16.2	16.4	16.5	16.1	0.38	2.36	16.1 ± 0.40
5.	20	19.5	19.8	19.9	20.4	20.8	20.5	20.1	0.50	2.47	20.1 ± 0.52
6.	24	23.9	23.8	24.1	24.2	24.4	24.5	24.2	0.28	1.16	24.2 ± 0.29
7.	28	27.9	27.5	27.9	28.2	28.5	28.5	28.1	0.39	1.39	28.1 ± 0.41
8.	32	31.8	32.0	32.0	32.3	32.6	32.1	32.1	0.31	0.96	32.1 ± 0.33
9.	36	35.8	35.9	36.7	36.2	36.4	36.5	36.3	0.35	0.96	36.3 ± 0.37
10.	40	39.9	39.5	39.9	40.2	40.5	40.5	40.1	0.39	0.97	40.1 ± 0.41
11.	48	47.8	47.5	47.8	48.5	48.3	48.4	48.1	0.41	0.85	48.1 ± 0.43
12.	60	59.9	59.8	59.8	60.2	60.3	60.5	60.1	0.29	0.48	60.1 ± 0.30
13.	80	79.5	79.5	79.9	80.8	80.4	80.5	80.1	0.55	0.69	80.0 ± 0.58

Average of six determinations. Confidence limit = Mean ± ts/ N, where s is the standard deviation. N are the replicate measurements and t is the statistical factor that depends on the number of degrees of freedom at 95% confidence level and for five degrees of freedom (N-1) t = 2.571.

Further, a need exist for an improved, versatile spectrophotometric method for determining uranium in ores. This method should be rapid, simple, precise and applicable without modification to all types of ores with U_3O_8 contents as low as 0.01% which is probably the limit of commercial interest.

The chemistry of uranium makes it particularly adoptable to solvent-extraction separations. The best known solvent-extraction systems for uranium, are triisooctyl amine (TIOA), TOPO, and tributyl phosphate (TBP), but they have certain disadvantages. In the TIOA system, the uranium must be in a chloride solution which is not usual reactor fuel dissolution medium. In the TBP system, uranium is extracted from the more common nitrate solution, but the TBP must be diluted with an inert solvent to prevent extraction of other elements such as the lanthanides, niobium, and zirconium. Since dilution decreases the extraction coefficient of uranium as many as four extraction with diluted TBP may be required for complete recovery of uranium. The TBP system is also subject to interferences from phosphate and large quantities of sulfate both substances inhibit the extraction of uranium.

3.14 Application of U(VI)-DBM-3,5diBr PADAP Extractive Spectrophotometric Method in Presence of Various likely Associated Elements

The U(VI)-DBM-3,5 diBr PADAP method is highly selective when the extraction is carried out in presence of masking mixture of EDTA and sodium fluoride and salting-out agent mixture calcium nitrate and aluminium nitrate. The co-extraction of many associated elements is prevented in presence of the masking mixture in the initial extraction of uranium(VI) with DBM. However, anions like arsenate, carbonate, phosphate, tartrate and citrate have low tolerance limits while the interference of thorium persists even at low concentrations because of co-extractions of thorium.

3.14.1 *Analysis of Uranium in Water Samples*

While this method can tolerate upto 50 mg. of chloride, higher concentrations only decrease the sensitivity slightly which can be overcome by using standard Beer's law curve in that particular chloride concentration. To test the applicability of the present method for the estimation of uranium in various types of water samples of varying salinity, known amounts of uranium were spiked to different water samples and the uranium content was analysed adopting the procedure given in table 3.3 and the corresponding data is given in Table 3.8.

TABLE 3.8
Analysis of Uranium in Water Samples by U(VI)-DBM-3,5 diBr PADAP Method

*Sample**	*Sample Volume ml.*	*U(VI) added* μg	*U(VI) found* μg^{+}	*Recovery* %
	50.0	0.0	0.0	—
Reservoir	50.0	5.0	4.9	98
water	50.0	8.0	7.6	95
	50.0	10.0	9.9	99
	50.0	0.0	0.2	—
Water	50.0	5.0	5.0	96
Waste	50.0	10.0	9.8	96
	50.0	15.0	14.5	95
	15.0	0.0	3.6	—
Well Water	15.0	5.0	8.5	98
Water	15.0	15.0	18.6	100
	15.0	25.0	28.6	100

* The samples were preliminary digested with nitric acid
+ Mean of two parallel determinations

3.14.2 *Analysis of Uranium in Diverse Matrix Materials*

To revel the applicability of the present method, for the analysis of diversified matrix materials, uranium(VI) (100 µg.) is estimated in presence of likely associated metals adopting the U(VI)-DBM-3,5 diBr PADAP method and the corresponding data are given in Table 3.9.

3.14.3 *Analysis of Uranium in Presence of Large Concentration of Thorium*

Thorium interferes seriously in the present method when it is present in large quantities because of coextration of Th (IV) with DBM even in presence of masking agents. However, detailed investigations revel that interference of thorium can be removed by carrying out a preextraction of uranium with ethylacetate in presence of EDTA and aluminium nitrate and back-extracting any thorium co-extracted by scrubbing the ethylacetate extract with EDTA and aluminium nitrate. In the process of scrubbing with EDTA and aluminium nitrate, it is observed that thorium is preferentialy removed from ethylacetate extract into the aqueous phase while uranium remains in the organic phase. The uranium in the organic phase can then be determined preciseiy after evaporating the extract and following the U(VI)-DBM-3,5 di Br PADAP method.

TABLE 3.9

The Determination of Uranium in the Presence of Pb, Bi, Cu, As, Sn, Al, Fe, Ni, Co, Cr, W, Ti, Th, Be, Ce, Mg,

(100 μg of Uranium Introduced)

Extraneous elements mg.	*U found μg.*	*Relative error %*	*Extraneous elements mg.*	*U found μg.*	*Relative error*
Pb. 5	100.84	0.84	Cr 100	103.1	3.1
Pb. 50	100.6	0.6	W 30	103.1	3.1
Bi 20	100.84	0.84	W 50	103.0	3.0
Bi 100	100.6	0.6	Ti 5	103.5	3.5
Cu 10	100.0	0.00	Ti 20 (air was blown through)	104.2	4.2
Cu 100	100.72	0.72	Ti 20 (air was blown through)	100.4	0.4
Al 5	100.84	0.84	Th 10	100.2	0.2
Al 50	100.5	0.5	Th 100	102.1	0.1
Fe 20	101.5	1.5	Be 20	104.2	4.2
Fe 20 (tartaric acid added)	103.1	3.1	Be 40	103.5	3.5
Fe 20 F added	104.4	4.4	Ca 20	103.5	3.5
Ni. 40	104.2	4.2	Mg 40	103.5	3.5
Ni 100	103.5	3.5			
Co 50	103.5	3.5			
Co 100	103.0	3.0			

Preparation of Scrubbing Solution

Aluminium Nitrate Solution - Dissolves 1.0 kg. of aluminium nitrate Al $(No_3)_3$. $9H_2O$ in 500 ml distilled water by heating.

Tetrasodium EDTA Solution - Dissolve 25 g of disodium EDTA by slowly adding to it 134 ml of 1.0 sodium hydroxide solution, and dilute to 250 ml with distilled water.

EDTA - Aluminium Nitrate Solution - (1+3 v/v) - Add 25 ml of the above EDTA solution to 75 ml of aluminium nitrate solution and adjust the pH to between 1.4 and 1.7. There should be minimum change in volume while adjusting the pH. Which should not exceed 5 percent. This mixture should be prepared freshly before use, as EDTA is likely to separate out on standing.

Procedure

Weigh 1.0 g of thorium nitrate sample, transfer it to a 20 ml Pyrex beaker and dissolve it in 2.5 ml of 10 percent V/V nitric acid. Transfer the solution to 125 ml. separating funnel and add 50-200 mg of uranium(VI) to it. Then add 15 ml. of aluminium nitrate solution to the beaker and transfer it to the funnel. Add exactly 10 ml. of ethylacetate to the funnel, stopper and shake it for two minutes. Allow the layers to separate and drain off the aqueous phase. Scrubb the organic phase for 1 minute with 5 ml of EDTA- aluminium nitrate scrub solution, maintaining the pH at 1.5 and reject the scrubbing. Repeat this 3 or more times and finally evaporate the ethylacetate extract by warming on hot water bath. Dissolve the residue in 2 ml of 1:1 nitric acid and estimate the uranium in it U(VI)-DBM-3,5 diBr PADAP method given in table 3.3.

50-200 μg of uranium in presence of 1 g of thorium nitrate are estimated by following the above procedures and the corresponding values are presented in table 3.10

TABLE 3.10

Analysis of Uranium in Presence of Thorium Nitrate (ig)

S. No.	*uranium added μg*	*uranium found μg*	*Mean*
1.	Nil	0.04, 0.06, 0.05, 0.06	0.05 ± 0.01
2.	50	54.5,53.5, 50.5, 51.5	52.5 ± 2.0
3.	100	105.5, 107.5, 106.5, 102.5	105 ± 2.5
4.	200	208, 205, 204, 207	206 ± 2.0

3.15 Determination of Uranium in Minerals, Ores and Rocks and Soils

Methods of Decomposition

The process of extracting and dissolving uranium from uranium minerals and ores is a very important stage in the quantitative determination of uranium. This process is complicated by the fact that in uranium ores, and also in minerals containing uranium, there are considerably greater amounts of minerals of other elements.

It is well known that minerals are subdivided into primary and secondary according to their origin. Primary minerals are chiefly compounds of the oxides of uranium (IV), titanium, iron, tantalum, niobium, rare earth elements etc. Among these are uraninite and all its varieties, pitchblende, etc; all the primary uranium minerals are readily soluble in dilute and concentrated HNO_3,, as well as in HCl and H_2SO_4 in the presence of an oxidant.

Secondary minerals were formed by changes in primary minerals or their decomposition. They are phosphates, carbonates, arsenates, sulfates, vanadates, etc., containing uranium(VI); the most widespread of these uranium blacks (certain uranium-oxide minerals) autunite, uranophane, carnotite, schroeckingerite, Kasolite, torbernite and tyuyamunite. Secondary minerals are also readily soluble in minerals acids.

The most frequently encountered uranium ores are the hydroalumosilicate ores, with over 50% of SiO_2 and the carbonate-hematite-magnetite ores, containing large mounts of iron (upto 70%); rarer are coals containing uranium with over 50% of carbon and limestones, with considerable quantities (Over 40%) of calcium carbonate. The uranium in the ores is present chiefly as pitchblende, uraninite, torbernite, uranophane other minerals are rarer. Besides uranium the ores may also contain Ti, Al, Fe, Mn, Mg, Ca, Na, K, Cr, P, V, W, S, Bi, Pb, Zn, Cu, Co, Ni, As and more rarely Zr, Nb, Ta, etc.

The choice of the method for the decomposition of an ore or rock depends upon the chemical composition of the sample, as well as upon the distribution of uranium in the rock. A necessary condition for a rapid and quantitative extraction of the uranium into solution is that the ore or rock be first ground usually into a fine powder (150 to 200 mesh). The choice of the sample of an ore or a rock depends on the uranium content in the specimen as well as on the method to be used for the determination of the uranium (the usual samples are 0.2 to 3 g).

Ores containing organic matter must be ignited at 500-700° before the decomposition; ignition at very high temperature (1000-1100°)

is not recommended, because this makes the extraction of the uranium into solution more difficult. Usually uranium ores are decomposed by strong acids: HNO_3 HCI, HBr, $HCIO_4$. H_2SO_4 or their mixtures in the presence of an oxidant: in most cases the extraction of the uranium into solution is quantitative.

Ores with high silicate content are decomposed by hydrofluoric acid; however, some silicates are not completely decomposed by this method (e.g. zircon).

The residue remaining after the decomposition of the ore by strong acids and hydrofluoric acid contains Zr, Th, Ti, etc., and can be dissolved by fusing with alkalis, carbonates, sodium peroxide, bisulfate, and persulfate, acid fluorides, etc.

3.16 Procedure for the Analysis of Ore/Rock/Soil Sample

Ignite a suitable sample at 500-600° and then heat with conc-HNO_3 or aqua regia. Filter the undecomposed residue and wash. Ash and ignite the filter at 500-600°c. Treat the ignitied residue with concentrated hydrofluoric acid with some nitric acid. Repeat the hydrofluoric acid treatment until the silicates are completely decomposed. Since the presence of large concentration of F^- ion during further separation and determination of uranium is usually undesirable first decompose the ores with hydrofluoric acid and transform the fluorides into nitrates, by evaporating with HNO_3. Dissolve salts obtained in dilute HNO_3. Filter off any insoluble residue and then decompose by fusing it with potassium bisulphate.

The above procedure has been tested by analysing 50-100 µg of uranium in presence of various likely associated elements with simulated composition of uranium ores and rocks by U(VI)-DBM-3,5 diBr PADAP method and the corresponding results are given in Table 3.11.

Adopting this procedures, certain actual ore samples or Jagudada ore singhbhum thrust belt, Bihar, India and certain soil samples are also analysed and the corresponding results are given in Table 3.12 A and B.

The results obtained in these investigations indicate that the present method can be conveniently applied for the precise analysis of uranium at trace level in various types of uranium ores, rocks and soil samples.

Table 3.11

Determination of Uranium in Synthetic Samples Corresponding to the Composition Similar to Uranium Ores.

S. No.	Composition (gm) Al_2O_3	MgO	CaO	SiO_2	Fe_2O_3	Composition (μg) V	Co	Ni	Cu	Uranium (μg) Added	Found *	Recovery %	S.D.
1.	0.50	0.00	0.00	0.00	0.00	0	400	0	150	100	97.4	97.4	0.02
2.	0.15	0.06	0.07	0.00	0.00	200	0	300	0	50	48.5	97.0	0.01
3.	0.15	0.06	0.07	0.50	0.00	0	300	0	200	100	100	98.2	0.01
4.	0.15	0.06	0.07	0.50	0.01	100	0	200	0	100	97.1	97.1	0.04
5.	0.15	0.06	0.07	0.50	0.01	0	200	0	50	100	98.3	98.3	0.03
6.	0.15	0.06	0.07	0.50	0.01	0	0	0	0	100	102.0	10.2	0.04
7.	0.15	0.06	0.07	0.50	0.01	150	0	50	0	50	49.2	98.4	0.04
8.	0.15	0.06	0.07	0.50	0.01	0	100	0	100	50	48.5	97.0	0.02

* Average of three determinations.

Table 3.12 A

Results of The Determination of Uranium in Soils

No. of Sample	Uranium content, 1% 1st analysis	2nd analysis
1.	2.9×10^{-4}	3.0×10^{-4}
2.	3.5×10^{-4}	2.1×10^{-4}
3.	2.8×10^{-4}	2.5×10^{-4}
4.	1.9×10^{-4}	1.0×10^{-4}
5.	2.0×10^{-4}	2.0×10^{-4}
6.	2.1×10^{-4}	2.1×10^{-4}
7.	2.0×10^{-4}	1.9×10^{-4}

Table 3.12 B

Results of Determination of Uranium Expressed as U_3O_8 in Jagudada Ore, Sighbhum Thrust, Bihar, India Ore Sample

S. No.	*U_3O_8 %*	
	1st analysis	*2nd analysis*
1.	0.032	0.035
2.	0.045	0.048
3.	0.090	0.095
4.	0.060	0.065

3.17 Determination of Uranium in Alloys

Methods of Dissolving Uranium Alloys

The methods used to dissolve uranium alloys are mainly chosen by taking into account the nature of the alloying metal and the method to be employed in the further determination of the components.

As a rule, uranium alloys with different metals, copper, zinc, bismuth, etc, which dissolve fairly readily in mineral acids or their mixtures; the alloys with lead and bismuth react even with water[38].

Nitric acid or aqua regia are commonly used to dissolve uranium alloys; if the alloys contain Zr and Nb, small amounts of F^- ions are added to the HNO_3.

Many alloys can be dissolved in hydrochloric acid with the addition of an oxidant, and also in perchloric sulfuric acids.

Alloys of uranium with niobium and molybdenum are sometimes decomposed by a solution of caustic soda and hydrogen peroxide (by heating).

3.18 Separation of Uranium and its Determination in Alloys

Uranium-Bismuth Alloys

Uranium-bismuth alloys are usually dissolved in nitric acid. Milner and Nunn[39] used the separation on ion exchangers for the determination of uranium in ternary alloys with a bismuth base (U-Th-

Bi alloys); the uranium was sorbed together with the bismuth from a 5M HCl solution on a deacidite, FF ion exchanger in its chloride form. After the uranium has been eluted from the column by a 0.2 M HCl solution, it is reduced to uranium (IV) and titrated with cerium (IV) sulfate, or it is determined polarographically in a titrate electrolyte. The method gives satisfactory results, but is rather long as it is necessary to remove organic matter which is extracted from the anion-exchange resin and applicable at larger concentrations of uranium. Milner used similar method for the determination of uranium in a binary alloy with Bi[40,41].

In the analysis of ternary alloys with a bismuth base containing 0.2-5% of uranium and 0.2-5% of neodymium or praseodymium, Milner and Edwards[42] separated the bismuth by extraction with a chloroform solution of diethylammonium diethyldithiocarbamate from a solution of 1 N HNO_3, then the same reagent was used to extract uranium at pH 5.5-6.0 (with n addition of acetate buffer). The quantitative determination of uranium was carried out by the photometry of a chloroform solution of the diethyldithiocarbamate uranium complex.

Kirby and Grawlly[43] separated the uranium from bismuth, as well as from small amounts of Zirconium, magnesium, and iron contained in the alloy, by extracting the hydroxyquinoline-uranium complex with chloroform in the presence of complexone III as a masking agent.

3.19 Analysis of Uranium Alloys

A. Uranium-Zinc-Aluminium Alloys Procedure for Dissolution.

Dissolve 0.4 g of the alloy in 10 ml of Hcl with addition of H_2O_2 and evaporate the solution to moist salt. After cooling dissolve the residue in 1:1 HNO_3 and make up to 100 ml in a volumetric flask. Transfer an aliquot of the above solution and estimate uranium by U(VI)-DBM-3,5 diBr PADAP method.

Since U-Zn, U-Al alloys are not available, known concentration of uranium are spiked into 6% Zn-Al alloys (BCS-CRM no. 300/1) and U-Al-Mg alloy(BCS No. 316) and the corresponding results are shown in the Table 3.13

B. Uranium-Cooper-Bismuth Alloy :

Dissolve 0.2 g. of the alloy in 2 ml. of conc. HNO_3 and evaporate the solution to moist Salt. After cooling dissolve the residue in 1:1 HNO_3 and make up to 100 ml in a volumetric flask. Transfer an aliquot of the

Table 3.13
Analysis of Uranium Alloys

Alloy	Composition											Uranium spiked	Uranium found
	Si	*Fe*	*Cu*	*Mn*	*Mg*	*Cr*	*Zn*	*Ti*	*Zr*	*Pb*	*Ni*	*Sn*	*µg*	*µg*
1. u-Zm Alloy BCS No. 300 / 1	0.14	0.24	1.27	0.33	2.74	0.13	5.87	0.09	0.18	0.00	0.00	–	200	208 µg
2. u-Al-Mg Alloy BCS No. 316	0.055	0.009	0.04	0.28	–	–	0.68	–	–	0.24	0.04	0.005	200	210 µg
3. u-Cu-Al Alloy BCS No. 181 / 3	0.30	0.72	2.48	1.10	1.57	0.04	2.52	0.058	0.00	0.101	2.0	–	200	215 µg
4. 20% u-80% Bi alloy	200 mg of uranium + 800 mg or bismuth												200 mg. (1 g sample)	209 mg
													200 mg. (1 g sample)	204 mg

above solution and estimate uranium by U(VI)-DBM-3,5-diBr PADAP method.

In actual analysis for U-Cu alloy, known amounts of uranium are spiked into BCS-181/3 Cu-Al alloy and then uranium in it is estimated.

For U-Bi alloy, synthetic simulated mixture of uranium and bismuth in 20 to 80% respectively is taken and uranium content in it is estimated and the corresponding results are shown in Table 3.13

C. Uranium-Zinc Alloys :

Dissolve the alloy (0.04 g.) in 10 ml of HCl with the addition of hydrogen peroxide, dilute the solution, with water to 200 ml and then analyse an aliquot containing about 25 mg of uranium by U(VI)-DBM-3,5 diBr PADAP method.

D. Uranium-Aluminium Alloys

Dissolve 2.5 g. of the alloy, containing up to 20% of uranium in 50 ml of concentrated HCl and dilute with water to 100 ml and analyse uranium by U(VI)-DBM-3,5 diBr PADAP method.

E. Uranium-Zirconium Alloys

Dissolve 1.0 gm. uranium-zirconium alloys in 5.0 ml of conc. Nitric acid and 10.0 ml 5% NaF and heated in a nickel crucible. Repeat this process until a clean solution is obtained. After cooling make up the solution to 100 ml and estimate uranium by U(VI)-DBM-3,5 diBr PADAP method.

References

1. E.B. Sandell, *Colorimetric Determination of Traces of Metals*, 3rd ed., Interscience, New York, 1959, p. 915
2. J.S. Fritz and M.J. Richard, *Anal. Chim. Acta*, 28, 1959; 167.
3. H.P. Holcomb and J.H. Yoe, *Anal. Chem.*, 32, 1960; 612.
4. S.B. Savvin, *Talanta* 8, 1961; 673.
5. T. Taketatsu, M. Kaneko, and N. Kono, *Talanta*, 21; 1974; 87
6. K. Motojima, H. Yoshida, and K. Izawa, *Anal. Chem.* 32, 1960, 1083.
7. Om Prakash, S. Kumar and S.P. Mushran, *Talanta* 26, 1979; 1167.
8. S. Kumar, Om Prakash and S.P. Mushran, *Analyst* 105, 1980; 474.

9. K. Uesugi, *Bull. Chem. Soc.*, Japan, 42, 1969; 2998.

10. K. Uesugi, *Bull. Chem. Soc*, Japan, 42, 1969; 2051.

11. K. Uesugi and T. Shigematsu, *Anal. Chim. Acta*, 84, 1976; 377.

12. K. Uesugi and M. Miyawaki, *Microchem.* J., 26, 1981; 288.

13. Katsuya Uesugi and Tohru Nagahiro, *Anal. Chim. Acta*, 148, 1983; 315.

14. J. Korkisch and F. Hecht, *Hand Book of Anal. Chem. of Uranium, Quantitative Analysis*, Vol. VI b β-Springer-Verlag, Berlin, 1972.

15. J. Horak and A. Okac, *Collection Czech, Chem. Commun.*, 1968; 33, 304.

16. O. Prakash, J.R. Verma and S.P. Mushran, *Chim. Anal.* Paris, 1970, 52, 655.

17. L. Sommer, V.M. Ivanov and H. Novotna, *Talanta*, 1967, 14, 329.

18. E. Gagliardi and Ilmaier, Mikrochim. Acta, 1968, 1259.

19. L. Sommer, *Collection Czech, Chem., Commun.* 1970, 35, 794.

20. S.N. Sinha and A.K. Dey, *Indian J Chem.*, 1963, 1, 70.

21. F.J. Hurst, D.J. Course, and K.B. Brown, *Eng., Chem., Process Des. Dev*, 1072, 11, 122.

22. F.J. Hurst, W.D. Arnold and A.D. Ryon, *Chem. Eng.*, 1977, 84, 56.

23. B.B. Jablonski, and D.E. Leyden, Anal. Chem. 50, 404, 1978.

24. T.M. Florence and Y. Farrar, *Anal. Chem.*, 35, 1613, 1963.

25. T.M. Florence, D.A. Johnson and Y. J. Farrar, *Anal. Chem.*, 41, 1652 ; 1969

26. T.M. Florence and Y .Farrar, *Anal. Chem.* 42, 271;1970.

27. D.A. Johnson and T.M. Florence, *Anal. Chim. Acta*, 53, 73; 1971.

28. D.A. Johnson and T.M. Florence, *Talanta*, 22, 253; 1975.

29. P. Palkans, *Anal. Chim. Acta* 69, 211; 1974.

30. S.J. Lyle and M. Tamizi, *Anal. Chim. Acta* 108, 267; 1979.

31. Shui-Chieh Hung, Chang-Ling Qu and Shui-Sheng Wu, *Talanta*, 29, 629; 1982.

32. S.J. Lyle and M. Tamizi, *Anal. Chim. Acta* 121, 341; 1980.

33. T.P. Lynch, A.F. Taylor and J.N. Wilson, *Analyst, (London)* 108, 470; 1983.

34. E.B.T. Cook, *Report* 1841, *Council for Mineral Technology, Randburg*, 1976.

35. Ayres, G.H. Narang, B.D., *Anal. Chim. Acta*, 24, 241; 1961

36. Aryes, B.H. Anal. Chem., 155, 332; 1938.

37. Ringbom, A.I.Z., *Anal. Chem.*, 115, 113; 1938.

38. *Anal. Chem. of Uranium and Thorium*, Russian Translation (P.N. Palei, Editor); 1956

39. Shelton, R.A.J. Analyst, 82, 531; 1957.

40. Milner, G.W. and J.H. Nunn, *Anal. Chim. Acta*, 17, 494; 1957.

41. Milner, G. and G. Barnett, *Anal. Chim. Acta*, 17, 220; 1957.

42. Milner, G.W. and J.W. Edwards, *Anal. Chim. Acta*, 17, 259; 1957.

43. Aleksandrova, L.S., T.B. Gapon, N.F. Filatova, and K.V. Chumutov, Trudy Komissii, *Po Alaliticheskoi Khimi*, 6 (9), 235; 1955.

44. Kirby K. and R. Grawlly - *Anal. Chim. Acta*, 19, 363, 1958.

45. Milner, G.J.W. Edwards, *Anal. Chim. Acta*, 16, 109; 1957.

4

Results and Discussion of the Trace Analysis of Uranium

With rapidly expanding nuclear energy programmes and stringent environmental protection regulations for the control of uranium (VI) levels in the waste disposal, analytical studies on the chemistry of uranium has also been increasing by several orders of magnitude in the last two decades.

Extractive spectrophotometric methods play very important role in the uranium analysis of complex materials as they can be made very selective and sensitive by properly modifying and optimising experimental conditions such that they are suitable for the analysis of uranium in a specific matrix. By solvent extraction one can achieve preconcentration and selective separation of uranium from the matrix components. Further, by using a suitable combination of masking agents the selectivity, and using a chromogenic chelating dye like any pyridylazo compounds in surfactant medium the sensitivity can be increased many folds such that these methods can be adopted for the precise analysis at much lower concentrations in complex matrices.

Dibenzoylmethane(R) in its enol form at pH 6.5 to 8.5 forms a yellow uranyl chelate complex $U0_2R_2$ which can be used for extractive spectrophotometric determination of uranium.

The absorption maximum of the complex in ethyl acetate is at 395 nm, molar absorptivity 2.0×10^4 (= 0.088). Dibenzoylmethane itself absorbs at wave length shorter than 400 nm. Phosphate, citrate and most metals forming dibenzoylmethane complexes interfere. The uranium (VI) - chelates of dibenzoylmethane can be extracted by many water-immiscible organic solvents like chloroform, carbontetrachloride, benzene, toluene and xylene, which give coloured extracts.

It is observed in the author's investigation that the extracted faintly coloured U(VI) - DBM complex can be converted into an intensely coloured ternary complex by reaction with chromogenic complexing agents like 4 - (2 - pyridylazo) resorcinol (PAR), 1 - (2 - Pyridylazo) - 2 - napthol (PAN), 2 - (2 - pyridylazo) - 5 - diethyl - amino phenol (PADAP), 2 - (5 - bromo - 2 - pyridylazo) - 5- diethylamino phenol (Br - PADAP) and 2 - (3,5 - dibromo-2-pyridylazo) - 5 - diethyl - amino phenol (3,5 - diBr PADAP) of which 3,5 diBr PADAP is the best chromogenic reagent in presence of a surfactant medium like pyridine + alcohol which can offer micellar phase. These chromogenic complexing agents though can react with uranium (VI) in aqueous phase directly to produce coloured complexes, their analytical application are severely limited as many metal ions like Fe, Pb, Sn, Ag, Cu and Ni interfere seriously and require elaborate separation procedures. Further, the colour reactions are less sensitive in aqueous phase.

In view of the importance of the determination of Uranium (VI) at trace levels when associated with different interfering metal ions, the author has taken up a systematic study to develop a method for highly selective separation of uranium (VI) by extraction with dibenzoylmethane (DBM) into 15% pyridine + 10% n - butanol in chloroform mixed solvent from matrix components in presence of masking agent mixture (EDTA + Fluoride) followed by ternary complex formation in chloroform + n - butanol + pyridine medium, by reacting an aliquot of the addition of 2 - (3,5 - dibromo - 2 - pyridylazo) - 5 - diethylaminophenol solution in ethanol medium which produce intensely coloured species. This method is free from many interferences from a variety of diverse ions. To exploit this reaction system for the precise trace determination of uranium when present in various diversified complex matrix components, conditions are optimised based on systematic investigations.

Detailed and systematic studies are made for optimising experimental conditions for selective extraction of uranium using dibenzoylmethane and its sensitive estimation by interacting the extracted chelate system with a chromogenic chelating agent 3,5 - diBr PADAP in non-aqueous medium.

In presence of surfactant like pyridine detailed and systematic studies are made for developing highly selective and simple procedures for the extraction and preconcentration of uranium in various complex matrix components using suitable combination of masking agents and pH.

In the extraction of uranium with DBM, neutral donors like pyridine and n- butanol are found to exert synergistic effect, facilitating quantitative extraction of uranium. The extraction efficiency of various extracting solvents like carbontetrachloride,. chloroform, benzene, MIBK, ethyl acetate, isoamyl alcohol in the presence and absence of pyridine and n-butanol have been systematically investigated.

Though quantitative extraction can be achieved with carbontetrachloride, chloroform, benzene, nitróbenzene and MIBK in the presence of pyridine and n-butanol, 15% pyridine + 10% n butanol in chloroform is found to be the best extracting mixed solvent for the extraction of U(VI) - DBM system, keeping in view of its speedy attainment of equilibrium with intense colour development with 3,5, - diBr PADAP system.

Though there is a marginal increase in the extraction of U(VI) - DBM in presence of pyridine + n-butanol, it is observed that pyridine + n-butanol medium facilitates extraction of U(VI) in a single extraction. Further, the presence of pyridine +n-butanol is found to be essential for the colour reaction between U(VI) - DBM and chromogenic chelating agent 3,5 - diBr PADAP. So, 15% pyridine + 10% n-butanol in chloroform is used in all further investigations.

Variations in the volume of aqueous phase from 20 to 200 ml. do not effect much the extraction efficiency of U(VI) - DBM - pyridine + n-butanol in chloroform system. However, when the aqueous phase is above 100 ml longer equilibrium periods are required.

U(VI) - DBM system is found to require a shaking period of 2 minutes only for quantitative extraction of U(VI) - DBM with 15% pyridine + 10% butanol on chloroform.

Unlike other metals uranium (VI) does not form complexes with EDTA or CYDTA which can be used as masking agents for decreasing the interference of co-extracting metal ions. If EDTA is used along with magnesium chloride it enhances the extraction kinetics of uranium (VI) facilitating for a fast selective separation of uranium from co-existing matrix components.

Though EDTA, CYDTA can be used as good masking agents for suppressing the interference of several metal ions, in presence of these masking agents, the extraction of uranium becomes less quantitative with pure solvents requiring several extractions with increased equilibrium periods.

It is observed that the extraction of U(VI) with dibenzoylmethane in presence of masking agents like EDTA or CYDTA can be made more selective and quantitative by using a mixed solvent like pyridine in chloroform. The selectively extracted colourless U(VI) - DBM complex can be directly converted into an intensely coloured mixed ligand complex by reacting with 3,5 - di Br PADAP in pyridine + chloroform medium which can be made as a highly sensitive method for the precise estimation of uranium. This procedure can also be applied for the highly selective separation of U(VI) of 10-100 mg. from 1 gm. of the matrix components like thorium, iron, copper, molybdenum, vanadium, chromium and zirconium.

Uranium (VI) is known to be extracted in the pH range 1-10 with various solvents like carbon tetrachloride, chloroform and requires 2-3 extraction with fresh solvent for quantitative extraction, particularly in presence of masking agents. The extraction of U(VI) - DBM with 15% pyridine + 10% n - butanol in chloroform mixed solvent in presence and absence of masking agents is examined under varying pH conditions.

In the study, in presence of masking salting-out agents, extractions were carried out under varied pH in presence of 0.5 ml each of 15% EDTA and 2% sodium fluoride and 2.5 ml each of 1% calcium nitrate and 5% aluminium nitrate.

It is well known that while at pH values less than 2.5, uranium (VI) exists as soluble, uranyl ion ($U0_2^{2+}$), at pH values higher than 2.5 it undergoes hydrolysis forming predominate complex hydrolysed ions which precipitate and hinder the extraction of uranium. However, DBM undergoes protonation at pH less than 2.5, which is not suitable for extraction, while its neutral species which exists between 2.5-7.0 pH form extractable neutral chelates with uranium. In view of the above, quantitative extraction of uranium with dibenzoylmethane cannot be achieved in the absence of any masking mixture. In the presence of masking mixture like EDTA and sodium fluoride, which form weak soluble complexes with uranium and act as auxiliary ligands preventing the hydrolysis of uranium in their precipitation and the extraction of uranium in their presence becomes quantitative with dibenzoylmethane.

The hydrolysis of a uranyl ion is a complex phenomenon and much research work has been published [1-9].

The following hydrolysis products are assumed : $U_20_5^{2+}$, $U_30_8^{2+}$, $[U0_2(OH)]^+$, $U_30_8(OH)_2$, $[U_30_8(OH)_3]^-$ and others.

Polynuclear doubly charged uranium-oxygen ions have a complicated structure and the corresponding hydrolysis constants of uranyl ion are given in Table 4.1

Table 4.2 gives pH values at which hydroxide precipitation begins as a function of uranyl ion concentration.

TABLE 4.1
Hydrolysis Constants of $U0_2^{2+}$

Hydrolysis reaction		*Hydrolysis Constant*	*Reference*
$U0_2^{2+} + 2H_20$	⇌	$(2.0+0.4)10_{-5}$	1
$U0_2(OH)^+ + H_30^+$		6.4×10^{-5}	2
		6.0×10^{-5}	10
$U0_2(OH)^+ + 2H_20$	⇌	406×10^{-4}	10
$U0_2(OH)_2 + H_30^+$			
$2U0_2^{2+}+3H_20$	⇌	1.35×10^{-6}	10
$U0_3.U0_2^{2+}2H_30^+$			
		2×10^{-6}	10
		1.14×10^{-6}	9
		$10^{-6.05}+0.1$	6
		4.1×10^{-3}	1
$U_2O_5^{2+}+U02^{2+}3H_20$	⇌	5×10^{-9}	9
$U_30_8^{2+}+2H_30^+$			
$U_30_8^{2+}+2H_20$	⇌	2.8×10^{-4}	9
$U_30_8(OH)_2+H_30^+$			
$U_30_8(OH)^++2H_20$	⇌	$3.0 \times 10^{-}$	9
$U_30_8(OH)_2+H_30^+$			
$U_30_8(OH)_2+2H_20$	⇌	4.0×10^{-8}	9
$U_30_8(OH)_3^-+H_30$			
$U_30_8(OH)_3+2H_20$	⇌	1.0×10^{-11}	9
$U_30_8(OH)_4^{2-}+H_30^+$			
$U_30_8(OH)_4^{2-}+2H_20$	⇌	4×10^{-12}	9
$U_30_8(OH)_5^{3-}+H_30^+$			

TABLE 4.2
pH at the beginning of uranyl hydroxide precipitation as a function of the UO_2^{2+} concentration in solution

Concentration of uranyl perchlorate, M	0.1	0.01	0.001	0.0001	0.00003	0.00001
pH at the beginning of hydroxide precipitation	4.47	5.27	5.90	6.62	6.80	7.22

As can be seen from the table, above pH 7 as stronger composite hydrolysed ions of uranium predominate resulting in its precipitation, the extraction steeply falls, both in the presence and absence of masking agents. Thus in the extraction of uranium use of masking agent mixture like EDTA and sodium fluoride will be useful both for preventing hydrolysis of uranium between 2.5-7 pH and for suppressing the interference of other associated metal ions.

As EDTA and fluoride form soluble weak auxiliary complexes with uranium (VI) when compared with dibenzoylmethane they can be tolerated in large concentration in the extraction of uranium with dibenzoyl methane in the pH range 4 to 7. Further they prevent the hydrolysis of uranium and mask the interference of many other metal ions which form strong colourless complexes with EDTA and fluoride. Further, any excess of these masking agents can be removed by the addition of aluminium nitrate and calcium nitrate as salting-out agents.

The sulphate ion amounts upto 2 mg/ml have no appreciable effect on the completeness of the extraction of uranium. If sulphate is present in large amounts, its interference can be removed by using calcium nitrate as the salting-out agent. If phosphates and sulphates are present simultaneously, a salting-out agent which consists of a mixture of aluminium and calcium or iron and calcium nitrate can suppress their interference upto 0.5 mg.

So the interference of other anions and cations on the extraction of U(VI)-DBM-Pyridine system has been studied in presence of EDTA and sodium fluoride as masking agents and calcium nitrate and aluminium nitrate as salting-out agents.

Cd, Zn, Cr, Mg, Ba and Sb donot interfere even in the absence of masking mixture while the masking mixture of EDTA and sodium fluoride binds Hg, Pb, Cu, Sn, Pt, Fe, Ni, Co, Mn, Nb, and Zn and remove their interference in the extraction of uranium (VI), with U(VI)-DBM-Pyridine system. However, serious interference is caused by W, Th, arsenate, Carbonate, Phosphate, tartrate and citrate when present in large concentration.

The term separation factor may be used as an index of effectiveness of separation of metal ions which is one of the main objectives of solvent extraction. If extraction of more than 99% of one metal and co-extraction of less than 1% of another metal is desired by a single extraction process, then a separation factor higher than 10^4 is necessary. The extraction of U(VI) with DBM using 15% pyridine + 10% n-butanol in chloroform in presence of masking agent mixture and certain metal ions which are likely to encounter in the complex matrices in varied concentration has been examined. Very effective separation of U(VI) from a number of diverse metal ions is achieved with DBM into 15% pyridine + 10% n-butanol in chloroform in presence of masking agent mixture. The extraction of uranium (VI) into varying concentration of dibenzoyl methane into 15% pyridine + 10% n butanol in chloroform are studied and a ratio of 1:2 for uranium to dibenzoyl methane is found to exist in the extracting species. The extraction process can be expressed as follows :

$$[U0_2(x)_2]_{aq}+2[HDBM]_{org}+n[R]\ org \rightleftharpoons [UO_2(DBM)_2nR]$$

Where

X = auxiliary ligand,

HDBM = enol form of dibenzoyl methane,

R = neutral donor pyridine and n-butanol.

Pyridylazo compounds with a hydroxy group in the ortho position with respect to the azo group are valuable chromogenic reagents. They have been widely used in comlexometric and extraction photometric methods for the determination of many ions. Highly sensitive reagents have been obtained by the introduction of the diethylamino group, which is a strongly electron - donating group in the para position next to the azo group. The azo dyes 2-(2-pyridylazo)-5-diethyl amino phenol (PADAP) and its bromine substituted derivative 2-(3,5 di bromo -2-pyridylazo)-5-diethyl amino phenol (3,5 di Br PADAP), the latter having a greater tolerance to iron (III), offer as sensitive reagents for the spectro photo metric determination of metal ions.

It is observed by the author that when mixed solvent like pyridine in chloroform or pyridine in carbontetra chloride is used, the extraction of U(VI) with DBM becomes quantitative in a single extraction over a wider pH range and further in presence of calcium nitrate + aluminium nitrate + EDTA, the extraction becomes highly selective and quantitative. Further it is also observed by the author that the selectively extracted colourless uranium (VI) - DBM complex can be converted into an intensely coloured mixed ligand complex by reacting with a chelating chromogenic complexing agent like 3,5 di Br PADAP in a surfactant medium like pyridine in chloroform. Due to interaction of dye and surfactant in micellar medium, it is observed that the colour of uranium (VI) - DBM - 3,5 di Br PADAP system is sensitized further by increasing the colour sensitivity accompanied with red-shift.

It is observed that when 3,5 di Br PADAP in ethanol is used in the DMF as a diluent maximum absorbance is obtained for the U(VI) - DBM - 3,5 di Br PADAP system.

The effect of variation of concentration of 3,5 di Br PADAP on the absorbance of U(VI) - DBM - 3,5 di Br PADAP mixed ligand complex in ethyl alcohol medium is investigated using different quantities of 3,5 di Br PADAP. It is found that 2.0 ml x 10^{-4}M 3,5 di Br PADAP in ethyl alcohol is sufficient for attaining maximum absorbance.

Absorption Spectra

U(VI) - DBM - 3,5 di Br PADAP mixed ligand complex showed maximum absorbance at 590 nm and the reagent blank has no significant absorbance at 590 nm. U(VI) - DBM exhibits maximum absorbance at 400 nm when measured against its reagent blank.

The U(VI) - DBM - 3,5 - di Br PADAP system in 15% pyridine + 10% n-butanol in chloroform medium is found to obey Beer's law upto 7.0 μg/ml of uranium (VI).

The molar extinction coefficient (E) and specific absorptivity (a) for U(VI) - DBM - 3,5 di Br PADAP mixed ligand complex into 15% pyridine + 10% n-butanol in chloroform mixed solvent medium are found to be E = 9.1 x 10^4 l mol^{-1} cm^{-1} and a = 0.3824 ml g^{-1} cm^{-1}. The Sandell's sensitivity of the colour reaction is found to be s = 0.00262 μg/cm^2.

It is also observed from the Ayre's plot and Ringbom's plot that the optimum range with minimum photometric error is in the region 1.6 to 4.0 μg/ml of uranium (VI) with a layer depth of one cm.

The precision of the method is studied by analysing 50 ml sample solutions of uranium (VI) varying from 4-100 ppm. The average standard deviation and coefficient of variation values are found to be 0.38 and 2.1 respectively.

Analytical Applications

The U(VI) - DBM - 3,5 di Br PADAP method is highly selective when the extraction is carried out in presence of masking mixture of EDTA and sodium fluoride and salting - out agent mixture calcium nitrate and aluminium nitrate. The coextraction of many associated elements is prevented in presence of the masking mixture in the initial extraction of uranium (VI) with DBM. However anions like arsenate, carbonate phosphate, tartrate and citrate have low tolerance limits while the interference of thorium persists even at low concentrations because of co-extraction of thorium.

Analysis of Uranium in Water Samples

While this method can tolerate upto 50 mg of chloride, higher concentrations only decrease the sensitivity slightly which can be overcome by using the standard Beer's law curve constructed in that particular chloride concentration. Detailed investigations reveal that the present method can be applied for the precise estimation of uranium in various types of water samples of varying salinity.

The applicability of the present method has also been demonstrated for the analysis of diversified matrix materials by actually analysing uranium (VI) (100 ug) in presence of likely associated metals like Pb, Bi, Cu, As, Sn, Al, Fe, Ni, Co, Cr, W, Ti, Th, Be, Ca, Mg, in the range 5-100 mg.

Thorium interferes seriously in the present method when it is present in large quantities because of co-extraction of Th(IV) with DBM even in presence of masking agents. However, detailed investigations reveal that interference of thorium can be removed by carrying out pre-extraction of uranium with ethyl acetate in presence of EDTA and aluminium nitrate and back extracting any thorium co-extracted by scrubbing the ethyl acetate extract with EDTA and aluminium nitrate. In the process of scrubbing with EDTA and aluminium nitrate, it is observed that thorium is preferentially removed from ethyl acetate extract into the aqueous phase while uranium remains in the organic phase. The uranium in the organic phase can then be determined precisely after evaporating the extract and following the U(VI) - DBM - 3.5 di Br PADAP method. 50-200 μg of uranium in presence of ig of thorium nitrate are estimated by following the above procedure.

Determination of Uranium in Minerals\Ores\Rocks\Soils and Alloys

To test the applicability of U(VI)-DBM - 3,5 di Br PADAP method for the analysis of low concentration of uranium in minerals, ores, rocks, soils and alloys, procedures have been optimised for the dissolution of the above samples which are adoptable for the U(VI) - DBM - 3,5 di Br PADAP extractive spectrophotometric method. Wherever real samples are available, they were analysed by adopting these procedures. Wherever real samples are not available, uranium was spiked into synthetic mixture of simulated composition and then uranium is analysed after adopting the dissolution procedure and estimation procedure.

In the case of uranium alloys, as the real samples are not available with the author, known concentration of uranium was spiked into BCS alloys of Cu-Al, Cu-Zn-Al, Mg -Al and then analysed adopting the dissolution procedure and extractive spectrophotometric procedure. The above procedure has been tested by analysing 50 - 100 μg of uranium in presence of various likely associated elements with simulated composition of uranium ores and rocks by U(VI) - DBM - 3,5 di Br PADAP method.

The Optimised dissolution procedures have also been tested by U(VI)-DBM - 3,5 di Br PADAP method by actually analysing ore samples of Jagudada ore Singhbhum Thrust belt, Bihar, India and certain soil samples also, which gave comparable and acceptable results.

The results obtained in these investigations indicate that the present method can be conveniently applied for the precise analysis of uranium at trace level in various types of uranium alloys, ores, rocks and soil samples.

To reveal the status of the present method a comparison of optimum experimental conditions and salient features of U(VI) - DBM - 3,5 di Br PADAP extractive spectrophotometric method along with other comparable methods are summerised in the Table 4.3.

TABLE 4.3
Comparison of Sensitive Spectrophotometric Methods for Uranium

S. No.	*System*	*Wavelength nm*	*Colour and conditions of reactions*	*Interfering substances*	*molar Extinctions coefficient*	*Ref.*
1.	UO_2^{2+}-dibenzoyl-methane	400	Bright yellow, pH6-8 water - alcohol solution or an organic solvent	V, Fe (III), Cu Mo, F^-, Po_4^{3-} and others	17810	11
2.	UO_2^{2+}-1-(2-pyri-dylazo)-2-naphthol	570	Bright red, 80% acetone+pyridine	F^-, Po_4^{3-} H_2O_2	23000	12
3.	UO_2^{2+}-arsenazo I	595	Blue, the reagent is pink (1) for UO_2^{2+}, pH=4.5-8; (2) for U^{4+},	Th, Zr, Ti, Ta, F^- PO_4^{3-}, SO_4^{2-}	22900	13
4.	UO_2^{2+}-arsenazo III	670	Emerald green, the reagent is pink (1) for UO_2^{2+} pH=1.0-3.0 (2) for U^{4+}, 0.01-10 NHCl.	Th, Zr, Fe (III) E REE	above 73000	14
5.	UO_2^{2+}-4-(2-Pyridy-lazo)-resorcinol	530	Red, the reagent is yellow. pH = 8	Many elements	21000	15
6.	UO_2^{2+}-3,5, diBr PADAP	590	Violet, the reagent is red; pH=1.5-2	arsenate, co_3^{2-}, PO_4^{3-}, tartrate, citrate, Th.	91000	

References

1. Komar, N.P. and Z.A. Tret Yak - *Zh Akh*, 10, 236, 1955.
2. Ahrland, S. - *Acta Chim, Scand*, 3,374, 1949.
3. Arden, T. - *J. Chem. Soc.* P. 299, 1949.
4. Francherro, *J. Compt. Rend.*, 227, 1367, 1948.
5. Hearne, J. And A. White - *J. Chem. Soc.* P. 2168, 1957.
6. Hietanen, S and Z. Sillen - *Acta Chim. Scand*, 8, 1607, 1954.
7. Sillen, L - *Recueil tray. Chim*. 75, 705, 1956.
8. Sutton, J. - *J. Inorg and Nucl. Chem.* 1.68, 1955.

9. Sutton, J. - *J. Chem. Soc. Suppl.* Iss. P. 275, 1949.

10. Seaborg, G.T. and J. Katz *The Actinide Elements* - New York, McGraw-Hill, 1954.

11. Palei and Novikov, 1959.

12. Cheng, K.L - *Anal. Chem.* 30, 1027, 1958.

13. Fritz, J.S. and M. Johnson - Richard - *Anal. Chim. Acta*, 28, 167, 1959.

14. Nemodruk, A.A., Yu. P. Novikov, A.M. Lukin and I.D. Kalinina - *Zn Akh*, 16, 180, 1961.

15. Pollard, F.H. P. Hanson, and W.J. Geard - *Anal. Chim. Acta.* 20, 26, 1959.

5

Review of the Earlier Work on the Solvent Extraction and Spectrophotometric Studies on Thorium IV

The use of thorium as raw material for the manufacture of nuclear fuel as well as in recently developed technical fields has stimulated research for extensive analytical rapport for the determination of thorium in various complex matrices.

Thorium minerals are most often encountered in igneous rocks, pegmatites, veins and alluvia of which the last are the most important industrial sources. Almost all thorium minerals are resistant to natural attack so that there is little tendency for the concentration of the metal in secondary minerals as a result of such attack. The most important thorium minerals are monazite, thorite and thrianite. The thorium ion has only one valency of state in aqueous solution and therefore undergoes no reactions involving valency change. The ion is also colourless, and relatively few characteristic colour reactions are known.

A number of qualitative colour reactions of thorium ion are known. The reactions are not however, always specific because the rare earth elements and Zirconium show similar reactions. The determination of thorium in isolation presents no particular difficulties but this may be prolonged and complicated in presence of associated elements. The majority of published methods involve prior removal of impurities and only a few may be applied directly.

Spectrophotometric methods continue to be important for the determination of Th in solution, in view of the poor sensitivity of the atomic absorption and atomic emission spectrometry which are rapidly replacing molecular absorption methods for elemental analysis. For the trace determination of thorium though a number of spectrophotometric methods are reported, many of them lack selectivity, and reproducibility even with slight variation in experimental conditions. Shankar Das[1] review the various spectrophotometric methods for thorium upto 1979 which consist of reactions of 54 reagents with molar absorptivities ranging from 0.01 to 13 X 10^4 l.mol.$^{-1}$ cm.$^{-1}$.

The author reviewed the spectrophotometric methods reported for thorium not covered in this report along with commonly practised methods.

Thoron 1[2] reacts with thorium (IV) in acid medium to yield a red, water soluble complex which forms the basis for spectrophotometric determination. A 0.2-0.3 M HCl medium is found to be optimum. The molar absorptivity of the complex in 0.25 M HCl is 1.7 X 10^4 l.mol.$^{-1}$ cm.$^{-1}$ at λ max 540 nm. Fluoride, oxalate, phosphate and sulphate interfered in this method while Fe (III) interference can be reduced with ascorbic acid.

Arsenazo III[3] reacts with thorium in strongly acidic solution. The reaction was very sensitive with molar absorptivity 1.15 x 105 l.mol.-1 cm.-1 at λ max 655 nm. Fluoride, phosphate and sulphate interfere in the reaction with arsenazo III. However, in this method even slight changes in acidity in blank and solution result in non-reproducible results.

O/m-/and P-nitrophenylazo chromotropic acids were prepared as colorimetric reagents for thorium by coupling diazotized nitroanilines with chromotropic acid and their utility for spectrophotometric study of the thorium complex was done by Kyoji et.al[4]. These reagents from 1:1 complex with metal in acidic solution and also in the pH range 5 to 12.

Spectrophotometric determination of thorium (IV) with bromopyrogallol red was studied at pH 5.7 (acetate buffer). Before measuring the absorbance, heating on water bath for about 20 minutes was required. The molar extinction was reported at 645 nm. Al, La Zr, Y, Cd, F^-, oxlate, tartrate, citrate, phosphate and EDTA were found to interfere.

The reaction of 4(2-thiazolylazo) resorcinol with thorium at pH 7.4 was studied[6] spectrophotometrically. Molar extinction was measured at 615 nm. The colour of the complex was reported as red-violet with ratio 1:3 between metal and reagent. The stability was reported as 2 hours. Beer's law was obeyed upto 95 μg of thorium per 24.0 ml. The molar extinction was reported as 3.15 x 10^4 l.mol.$^{-1}$ cm.$^{-1}$ while the sensitivity as 0.008 ug. per sq. cm.

The reaction of thorium with o-cresol phthalenon S, which is a component of xylenol orange was studied[7] at pH ranges 1.5-2.5 and at pH 3-6. The ratio between metal to reagent was reported as 1:1 at pH 1.5 -2.5 and 1:2 at pH 3-6. The colour of the complex was stable for 2-3 days which was developed instantaneously. The molar absorptivities were reported as 2.95 X 10^4 l.mol.$^{-1}$.cm.$^{-1}$ for 1:1 complex at pH 2.0 and 2.43 X 10^4 l.mol.$^{-1}$ cm.$^{-1}$ for 1:2 complex at pH 4.0. The absorption maxima were studied at 540 nm. The interference of Fe^{3+}, Ga^{2+}, Zr $(OH)_2^{2+}$, F^-, Po_4^{3-}, $WO^{2-}{}_4$ and $C_2 0_4^{2-}$ were studied. Beers law was obeyed for 1-9 μg\Th/ml.

Thorium was determined spectrophotometrically with arsenazo-III after extraction with thenoyl trifluoroacetone[8]. Thorium was separated from uranium. The earlier method of Khopkar and De[9] was examined for solutions containing or not containing oxalic acid. At pH one from aqueous solutions containing microgram amounts of thorium was extracted with 0.15 M 2-thenoyl trifluoroacetone. The shaking time was reported as 15.0 minutes. Thorium was stripped from organic phase with 3N HCl and determined at pH 1.5 spectrophotometrically using arsenazo III at 665 nm. At 650 nm uranium was determined with arsenazo III. The extraction of uranium was eliminated by extracting thorium at P^H 5.5 in presence of oxalic acid (0.1 to 0.3 gm per 25 ml) with thenoyl trifluoroacetone in benzene with shaking time about 20.0 minutes. Thorium was determined spectrophotometrically after separating zirconium and other rare earths through DOWEX 1-X8. The lower limit of detection was reported as 0.7 μg[10].

Spectrophotometric determination of thorium with carboxy arsenazo [3-(2 arsonophenylazo) -6-(2- carbonyl-phenylazo] chromotropic acid was reported[11]. The interference of Fe (III) was studied and was reduced with hydroxylamine hydrochloride. The molar extinction was recorded at 675 nm up to 150 μg. uranium 10 fold to thorium did not interfere whereas Zr and F^- interfered seriously.

Dragulescu et.al[12]. examined the absorption spectra for alizarin and quinalizarin and their complex with thorium in n-butanol. Thorium alizarin complex was obtained at pH 2.8 containing trichloroacetic acid. Thorium quinalizarin complex was obtained at pH 3.5. Both complexes were reported as having 1:1 composition. Al(III), Fe(III), U(VI), Zr(IV) and W(VI) interfered. Cu was masked with thiourea.

Thorium in rocks was estimated with arsenazo III by eliminating the interference of titanium by extracting it with alamine-336 in toluene from 9 M HCl medium[13]. The same procedure was used to eliminate the interference of iron and zirconium.

Thorium both at trace level and in large concentrations was determined[14] by separating it from all matrix components by anion exchange and with thoron or arsenazo III. The method was applied to geological samples.

Thorium-chlorophosphonazo-III complex was extracted in 3-methyl-1-butanol from 2-3 M hydrochloric acid solution, by T. Yamamoto[15]. Maximum absorbance was measured at 620 and 670 nm and Beer's law was obeyed at the latter wave length over the range of 0.15 µg/10 ml. of organic phase with molar absorptivity 12.2 X 10^4 $l.mol.^{-1} cm^{-1}$. Thorium was determined in presence of fluoride, oxalate, sulfate and EDTA. Zirconium, uranium, and niobium interfere seriously in this method.

The study of thorium separation by N-oxides of 5(4-pyridyl) nonane and trioctylamine from different mineral acid solutions by solvent extraction was studied by Ejaz[16]. The possible mechanism of extraction was studied, with extraction isotherms. A method of separation for thorium from rare earth elements and yttrium was done.

A single dissolution and single solvent extraction step for the spectrophotometric determination of thorium and uranium in ores was reported[17] where trioctyl phosphine oxide was used for the extraction.

Thorium in natural waters was determined by selectively extracting with aliguat 336 dissolved in xylene. It was determined spectrophotometrically with arsenazo III[18]. The method was finally applied to the analysis of Italian water samples.

The effects of certain hydroxy acids on the spectrophotometric determination of thorium (with alizarin S) and zirconium [with alizarin S or methylthymol blue] was studied by zolotokhin V.K.[19]. and showed that tartaric, citric, trihydroxyl, glutaric and gluconic acids each cause some decrease in the absorbance of solution of complexes of Zr or Th with the above reagents. By introducing the corresponding acid into calibration solution, correction was made.

Thorium was extracted in nitric acid medium with thenoyl trifluoroacetone in benzene and was estimated directly in the organic phase by spectrophotometry as its thoron complex and the interferences in this system were studied by Venkatesan et.al[20]. Interference of Fe(III) upto 5.0 mg was suppressed by adding ascorbic acid before extraction with thenoyl trifluoroacetone. Uranium upto a 2000 fold to thorium, did not interfere in this method while Zr interfered in this method.

At pH 4.9 aqueous aniline blue was added in presence of benzoate solution to thorium and its spectrophotometric determination was reported[21]. The absorption maximum was recorded at 660 nm with E=3160 $l.mol.^{-1}$ $cm.^{-1}$ The interference of Zr(IV), Mo(VI), Fe(III) and PO_4^{3-} was studied.

Thorium (0.5 -5g/1) was almost completely extracted from 0.1 M acetic acid at pH 4.7 into benzene solution of 16% versatic 911 which is a mixture of saturated tertiary monocarboxylic (C_9-C_{11}) acids[22]. The volume ratio of aqueous phase to organic phase can be increased upto 2:1 while standing time of 5 minutes is required for optimum extraction. Thorium was stripped back into 2M-HNO_3 from benzene solution. The interference of uranium, cerium and lead was critically studied.

Traces of thorium in soil samples was determined spectrophotometrically by using arsenazo III as the coloring reagent[23]. Rare-earths, Zr, Ti and U interfere seriously in the thorium determination and their interference was successfully eliminated with the addition of EDTA, triethanolamine, ascorbic acid and oxalic acid. 0.1 gm of soil sample was fused with $Na_2 0_2$ in a highly pure aluminium crucible (thorium free). The solution was boiled and filtered after leaching. The thorium hydroxide precipitate was washed with 1% hot NaOH, hot water and dissolved in hot 4 N HCl. A small quantity of solid ascorbic acid, one ml of 5% oxalic acid and one ml of 0.1% arsenazo III were added to this solution and diluted to 25 ml with 4 N HCl and thorium content was determined spectrophotometrically by measuring absorbance at 660 nm.

Yamani and Shabama[24], studied the extraction of thorium from nitrate solutions with quaternary ammonium halides. Thorium was separated from many elements normally present in monazite and fission products from nuclear fuel by extraction from 0.25 M HNO_3 -2M $NaNO_3$ medium into a 0.1 M solution of hyamine 1622 into dichloroethane. Thorium was then stripped from the organic phase with 0.1 M H_2SO_4 and determined complexometrically with xylenol orange as an indicator.

The reactions of Th(IV), U(VI) and U(VI) chromoazurol S, the effects of P^H and concentration on the composition and absorbance of the complexes were critically studied[25]. The optimum P^H range for complex formation was reported in acetate buffer medium of P^H 5 to 6. The λ max for thorium complex was reported as 550 nm while 550 to 600 nm for uranium complexes depending on the composition. To minimise the effect of hydrolysis the absorbance of U(VI) was measured after 5.0 minutes whereas for thorium it was reported that it was measured after 20.0 minutes.

Spectrophotometric determination of thorium in presence of lanthanides and uranium was reported[26] using the reaction between thorium and 4 (2,3,4 - trihydroxy phenylazo) benzene sulfonic acid. The metal : reagent ratio was reported as 1:2 in aqueous HCl at P^H 2.0 containing 80 micromoles of benzene sulfonic acid. Maximum absorption was reported at 420 nm with molar absorptivity 84,000 $l.mol.^{-1}$ $cm.^{-1}$. Beer's law was obeyed for 0.9 to 9.3 g/ml of thorium. 75000 fold amounts of lanthanides and 4000 fold amounts of U(VI) relative to thorium can be tolerated.

Jaiswal[27] extracted Th (20 uM) from HNO_3 CCl_4 solution of tri-isoamyl phosphate. Thorium was determined in both aqueous and organic phases with arsenazo III. HNO_3 was determined potentiometrically and water in the organic phase was determined by Karl Fischer reagent at pH 4.0 to 4.5 Th was almost quantitatively extracted with 18% of triisoamyl phosphate solution in carbon tetrachloride.

New colour reagent chlorophosphonazo-p^B was synthesized and its reaction was studied with thorium in presence of cationic surfactant by Zhang[28]. The reagent 3-4 (bromophenylazo) 6-4-chloro-2 phosphono phenylazo) chromotropic acid in HNO_3 or oxalic acid medium reacted with Th in the presence of hexadecyl trimethlyl ammonium bromide fo form a blue-violet complex with max at 685 nm (E 350,000 $l.\ mol.^{-1}$ $cm.^{-1}$). Beer's law was obeyed upto 0.2 μg/ml. of Th.

Thorium was extracted[29] in pentanol containing 0.1% NPBHA at pH 2 to 6.5. It was back extracted into 0.2 M HCl to which aqueous solution of 0.1% thorin was added and diluted to 10.0 ml. The absorbance maximum was reported at 545 nm. Thorium was measured in the range 0.5 to 12.5 ppm. Interference from Mo(IV), V(VI), W(VI) and Cu(II) was reported while Zr interference was eliminated with PO_4^{3-}.

The formation of stable purple-red complex between thorium and chlorophosphonazo -mN in phosphoric acid - oxalic acid-HNO_3 media was studied[30]. The molar absorptivity and Sandell sensitivity were reported as 8.23 X 10^4 l.mol.$^{-1}$ cm.$^{-1}$. and 2.82 X 10^{-3} μg/cm^2 respectively. The absorption maxima was reported at 670 nm. The method was useful in presence of large excess of uranium and iron. The lower detection limit was reported as 0.0001% Th. Beer's law was obeyed for 0-50 μg Th/25 ml.

The reaction between arsenazo-I and diazonium salt of 4 amino antipyrine producing 3(2-arsonophenyl) azo-6- (4-antipyryl) azo]-4,5-dihydroxy-2,7-naphthalene disulfonic acid and its reaction with thorium(IV) was studied[31]. The complex shows two absorption maxima at 645 and 655 nm. Beer's law was obeyed upto 4.3 μg Th(IV) λ max at 665 nm with molar absorptivity of 5.52 X 10^4 l.mol.$^{-1}$ cm.$^{-1}$. The complex possesses a 1:2 metal : ligand ratio.

The reaction of (1-2 thiazolylazo)-2 naphthol (TAN) with Th(IV) in presence of antipyrine was studied[32] and formation of sparingly soluble red chelate was reported. Glycine buffer was used to maintain pH 2.4 - 2.8 where complexation takes place immediately. Antipyrine was used to enhance the sensitivity of the complex which was stable for 19 hrs. 1:2 complex was reported between Th-TAN and maximum absorbance was reported at 555 nm. Beer's law was obeyed upto 6.56 g per ml with E=3.14 X 10^4 l.mol. $^{-1}$ cm.$^{-1}$. The Sandell's sensitivity was reported as 7.4 μg/cm^2.

The 1:2 complex of Th(IV) with the reagent 2,3,4- trihydroxy -2^1 carboxy -5^1 sulfoxo benzene in urotropin buffer at pH 3.0 was studied 33. The spectrophotometric determination was done at pH 3.0 and at the wavelength 490 nm Beer's law was obeyed upto 232 μg/25 ml.

A simple and convenient method for the direct titrimetric determination of thorium (IV) at micro gram level was developed[34] using. 1,2 - diamino cyclohexane - N,N,N^1 N^1 - tetracetic acid (DCTA) as a titrant and pyrocatechol violet (P V) as an indicator around 2.5-3.0. The optimum conditions like pH, concentration of DCTA, concentration of thorium, concentration of indicator, temperature and tolerance limits for foreign ions were also established. The method was used for the determination of thorium in monazite sand.

A new chromogenic reagent nitrophosphonazo-mN (NPA-mN)

2-(4-nitro-2-phosphonophenylazo) - 7 (3-nitrophenyl azo) - 1,8 -dihydroxy -3,6 -naphthalene disulfonic acid (NPA-mN) was used for the spectrophotometric determination of Th in strongly acidic medium[35] Beer's law was obeyed for 0-30 and 0-35 μg/25 ml. The relative standard deviation and coefficient of variation were 2.59×10^{-3} and 0.787% for 10 μg of Th. The method was applied for determination of Th in river water, urine and waste water samples.

1:1 complex with 3-nitro-α-methyl methane at pH 3-4 with Th(IV)-thorium (IV) was reported and thorium(IV) spectrophotometrically determined at 390 nm[36]. Beer's law was obeyed upto 5×10^{-5} M. Th. The molar absorptivity and Sandell's sensitivity were reported as 10,000 $l.mol.^{-1}$ $cm.^{-1}$ and 0.023 $\mu g/cm^2$ respectively.

Separation of thorium as ascorbate complex by extraction with aliquat-336 S was studied by Vibhute and Khopkar[37]. Thorium was quantitatively extracted into 0.1 M aliquat 336 S in xylene from 10 milli moleascorbate medium at pH 4.5 as aliquat 336 (S)- Thascorbate (1:1:1) complex and was back extracted into 2 molar HCl. No interference from Tl(I), Fe(II), As(III), Y and La, alkali metals or alkaline earth metals was reported. Ce(IV), V(V), Ti(IV) Mo(VI), Be(II), Sc(III), Zr(IV), U(VI) and Hf(IV) interfere as they were co-extracted with thorium. The method was applied for the determination of thorium in monazite and gas matles. The interfering elements were extracted and separated from organic phase by suitable aqueous phases. Thorium was back extracted into 2M - HCl and was determined by arsenazo (III).

Th(IV) was coprecipitated with LaF_3 followed by an extraction with 0.2 M TTA in benzene. The method was applied for environmental samples[38].

Simultaneous determination of uranium and thorium was reported[39] where they were extracted with trioctyl phosphine oxide in cyclohexane in presence of ascorbic acid, boric acid and aluminium salts. The solution was back extracted with 2N HNO_3 containing 0.8% NH_4F solution. Uranium was determined in the organic layer with 5-boromo-2-pyridylazo-5 diethylamino phenol spectrophotometrically. Thorium was determined in the aqueous solution with arsenazo III.

The other less familiar spectrophotometric methods of Th reported in the literature are briefly reviewed in Table 5.1 for quick revision.

TABLE 5.1

S. No.	Reagent	Conditions	Remarks	Ref.
1	2	3	4	5
(1)	Solochrome fast red	max 490 nm. The sample solution was treated with 2.5% ascorbic acid in methanol. HCl Medium.	Ti (IV), Zr (IV), Ce (IV), M (IV) interfered. The method was applied for the determination of thorium in marine sample. The molar absorptivity was 13,970 $l.mol.^{-1}cm.^{-1}$	41
(2)	3-methyl galangin	max 400 nm. The reaction was studied at PH	Metal to reagent ratio was 1:1 Beer's law was obeyed upto 13 ppm	42
(3)	bromopyrogallol red	The reaction studied at P^{H} 1-3 and P^{H}4-9 max 430 nm and 560 nm respectively	The molar absorptivities are 2.5=0.1 X 10^{4} $l.mol.^{-1}cm.^{-1}$ at P^{H} 2.0 and 6.3=0.2 X 10^{4} $l.mol.^{-1}cm.^{-1}$ at P^{H} 5.0 Depending on the OH the metal to ligand ratio changes.	43
(4)	N,N -Bis (2 hydroxy 5-sulfo phenyl)-C-cyanoformazon	The complex formation found to be affected by changing P^{R}. Inalkaline conditions the colour changes from orange red to blue.	Beer's law was obeyed upto 30 ppm. Molar absorptivity 39,000 $l.mol.^{-1}cm.^{-1}$.	44
(5)	Bis 2 ethyl hexyl Phosphoric acid	max 600 nm. A time gap of ten minutes was required before the measurement of absorbance	The molar absorptivity 4.9X10^{4} $l.mol.^{-1}cm.^{-1}$. Interference of U (IV), Se (IV) and Ti (IV) was observed.	45
(6)	anthrapurapur in	Two hours of standing time was required before going to measure the absorbance. max 540 nm	Beer's law obeyed upto 21 ppm.	46
(7)	Methylthymol blue	Th (IV) was extracted in to butanol at P^{H} 2.5-3.0 containing 0.05-0.1 M 1.2 diphenyl quanidine.	Molar absorptivity 6.6 X 10^{4} $l.mol.^{-1}$ $cm.^{-1}$	47
(8)	Chromozurol S	max 631 nm. The reaction was studied at P^{H} 5.6-6.1	The metal to ligand ratio was 1:4 V. No <Zr, W, Cu, Cr, Al interference Studied	48

(Contd.)

Table 5.1 (Contd.)

1	2	3	4	5
(9)	Schiff base derived from 1-amino-8-naphthol-1-3,6-disulfonic acid and salicyladehyde	The reaction studied at at P^{H} 3.2 to 5.2 max 450 nm	Molar absorptivity 7000 l.mo.$^{-1}$ cm Fe (III), Vo^{2-}_{3}, CrO^{2-}_{4} interfered	49
(10)	Querceiten	Optimum condition for complex formation P^{H} 4.0 to 7.0 max 425 nm	At P^{H} 3.0 max 541 nm fluorescence was studied at this wave length	50

References

1. Sankar Das. M., *International Union of Pure and Applied chemistry*, Vol. 55, 1231, 1983.
2. Palmer, A.R. *Anal. Chim. Acta.*, 14, 162, 1956.
3. Savvin, S.B. *Talanta* 8,673, 1961.
4. Kyoji Toei, Haruo Miyata, Hirosumi Kimura, *Bull. Chem. Soc.*, Japan, 40 (9), 2085, 1967.
5. Koichi Tonosaki, Takayoshi Sakai, *Bull. Chem. Soc.*, Japan, 4 (2), 456, 1969.
6. Koichi Tonosaki, Takayoshi Sakai, *Bull. Chem. Soc.*, Japan, 42 (9), 2719, 1969.
7. Cher Kesov A..I. Argisheva A.I., Konyakhina A.A., *Ftaleksony*, 173, 1970.
8. Hiroshi Onishi, Keichi Sekine, Japan *Analyst*, 19(4), 547, 1970.
9. Khopkar S.K. De *A.K. Analy. Ab.*, 8, 1114, 1961.
10. Michelle Archimband, Albannoel, Chi. *Analyst*. 53 (3), 161, 1971.
11. Kenj Motojima, Tadashi Yamamoto, Hiroshi Muto, Yoshiharu Kata, Japan *Analyst*. 19 (7), 935, 1970.
12. Dragulescu. D., Simonescu, T., Gizella Neme, Maria Totan, *Revue. Roum. Chim.* 15 (4), 563, 1970.
13. Pakalns P. *Anal. Chim. Acta*, 58, 463, 1972.
14. Korkisch, J., Dimitriadis, D., *Talanta* 20, 1199, 1973.
15. Yamamoto T. *Anal. Chim. Acta*, 63, 65, 1973.
16. Ejaz. M., *Talanta*, 23, 193, 1976.
17. Pollock E.N., *Analit. Chim. Acta*, 88, 399, 1977.
18. Cospito M. Rigali. L., *Anal. Chim. Acta*, 106, 385, 1979.
19. Zolotokhin, V.K., *Ukr. Khim. Zh.* 45 (11), 1118, 1979.
20. Venkatesan M., Gopalkrishnan V., Ramanjam A., Nadakarni M.N., India, *A.E.C. Bhabha Atomic Research Centre* (Re.) BARC, 1094, pp 9.1981.
21. Singh, N.S.B., Thankachan T.S., Balsubramanian G.R., *J. Radio Anal. Chim.* 60 (2), 395, 1980.

22. Ray Uday Sankar, Modak Sagarchandra, Sep. *Sci. Technolo.* 16 (1) 87, 1981.

23. Xuning Hu Yujan, Zhonghua Fangshe Yixue Y.U., *Fanghu Zazhi* 73, 50, 1982.

24. Abuzuhri A.Z. *Micro. Chem.* J. Jun, 29 (3), 345, 1984.

25. Yamami I.S.E.L., Shabana E.I. *Talanta*, 31, 627, 1984.

26. Tikhonov V.N. Lukina E.V. *Zh. Anal. Khim*, Sep. 39 (9), 1446, 1984.

27. Ganbarov, D.G., Ayubova A.M. Nguem Chichy *Zh. Anal. Khim.* 39 (8), 1440, 1984.

28. Jaiswal S., Rupain War D.C., *Acta. Chem.* Hung, 115 (3), 261, 1984.

29. Zhang N., Wu B., Qu. Y., Liu H., *Huaxue Shiji.*, 12 (2), 66, 1085.

30. Agrawal Y.K., Dayal U., *J. Radio and Nucl. Chem.* 90 (2), 303, 1985.

31. Yu Chengwei, Gran, Menggu, Liang, *Hong Fushe Fanghu* 5 (2), 112, 1985.

32. Aznarez Aldun J., *Lopez Molinero A.*, 3 (3), 234, 1984.

33. Sharma C.D. Eshwar M.c., *J. Radio and Nucl. Chem.* 91 (2), 323, 1985.

34. Gambarov D.G., Ayubova A.M., Yusubov N.N., Ng. Uygen Chichy. *Azerb. Khim. Zh.* 4, 131, 1984.

35. Maiwaluc, Srinivasulu K., *J. of Indian Chem. Soc.* Vol. XLI, 351, 1984.

36. Lin, Jummi, Gao Ming Xia, Liu, *Heng. Chen. Fenxi, Huaxue*, 12 (10), 900, 1984.

37. Bhanu Sabira, Tyagi M.D., Purohit D.N., *Orient. J. Chem.* 2 (2), 152, 1986.

38. Vibhute C.P. Khopkar S.M., *J. Radio and Nucl. Chem.* 97 (1), 3, 1986.

39. Pires do rio M.A. Godoy J.M., *Quim. Nova*, 9 (1), 98, 1986.

40. Ren Jiya, Zhang Jianquo, He, *Yafe Yanshi Kuanghu Ji Ceshi* 3 (3), 274, 1984.

41. Korkisch J., Janauer G.E., *Anal. Chem.* 33, 1930, 1961.

42. Katyal Mohan, Signh R.P., *Indian J. Chem.* 3 (6), 281, 1965.

43. Vaslenko V.D. Shanya M.V. Bolbas V.I. *Zh. Anal. Chem.* 22 (12), 1818, 1967.

44. Furukawa Masamichi, Sasaki Shozo, Goto Kazuo, Kamata Eijiro, Nakashima Ryozo, Shibita Shozo, Nagoya Kogyo Gijustsu, *Shikensho Hokoku* 17 (7), 161, 1968.

45. Cerrai E., Ghesini G. *Anal. Chim. Acta* 37 (3), 295, 1967.

46. Capitan F. Roman M. *Ar. Pharm.* 8 (7-10), 399, 1967.

47. Otomo Makoto, *Nippon Kagaku Zasshi*, 89 (5), 503, 1968.

48. Shijo Yoshio, Takeuchi Tsugio, *Bunseku Kagaku*, 18, 469, 1969.

49. Poddar S.N. *Indian J. Chem.* 3 (9), 407, 1965.

50. Babko A.K.. Chen Ty Hyeu, Volkova A.I., Getman T.E., *Ukr. Khim. Zh.* 35 (3), 292, 1969.

6

Studies on the Solvent Extraction of Thorium (IV)-TOPO-Complex Extracted into Toluene

6.1 Introduction

Thorium (IV) can be extracted from mineral acid solutions with various neutral extractants. Among them diethyl ether has long been reported to extract thorium (IV) from nitrate solutions by many workers[1-3]. The extraction, for example, 80% from the aqueous phase containing 3 M nitric acid and saturated with zinc nitrate[4].

Mesityl oxide[5-10] is one of the oxygen containing solvents which has been used for the extraction of Th from nitrate solutions. It extracts thorium (IV) quantitatively when the aqueous phase is acidified with nitric acid and saturated with aluminium nitrate[5]. Isomethyl butyl ketone[11-13] is also often used for this extraction. Cyclohexanone[14,15] and methyl cyclohexanone[16,17] have also been used for the separation of thorium by extraction.

An extractant much more effective is tributyl phosphate[18-24]. But dilution of TBP with an inert solvent decreases the extraction[11]. TBP extraction of thorium nitrate is better from acidified nitrate salt solution than from nitric acid at the same nitrate ion concentration[25-27], but the salt effect is different when the acidity is different[28].

From chloride solutions, with isomethyl butyl ketone the extraction of thorium (IV) is poor[11,13]. A systematic study of the extraction from ammonium thiocyanate solution with TBP in various diluents was reported. In presence of 1.0 M ammonium thiocyanate and 10% TBP with chloroform as diluent the extraction is poor whereas for inert solvent like carbon tetrachloride the extraction is practically quantitative[29].

Thorium in solutions of chloride, thiocyanate, nitrate and sulphate is extractable with large cations such as triphenylarsonium[30] tricaprylmethylammonium chloride[31-36] and several others[37-40].

The quadrivalent thorium is extractable effectively from strong acid solutions with diethyl hexyl phosphoric acid[41-43] and dibutyl phosphoric acid[44-46]. The extraction of thorium with acetylacetone is shown to be much more better[47-50,75,76]. Among diketones thenoyl trifluoro acetone (TTA) has been shown as an effective extractant[48-55]. Synergism was reported where the extraction of thorium is increased with TTA in presence of tributyl phosphate/trioctyl phosphine oxide/methyl isobutyl ketone[56-59]. Cupferron was also reported as an effective extractant[60-64]. N-phenyl n-benzo hydroxylamine was also reported as a good extractant[65-69,74]. In addition, there are many other extractants that have been used for the extraction of thorium (IV)[70-73].

Tri-n-octyl phosphine oxide which is a neutral organo-phosphorous compound is much more better extractant than tributyl phosphate. This reagent extracts thorium (IV) effectively even from nitric acid solution containing no other salt[11].

The use of thorium in the manufacture of nuclear fuel and in recently development fields show importance in its separation from other associated elements like calcium, magnesium, iron, aluminium tin, uranium and zirconium.

In the author's investigations it is observed that thorium (IV) can be quantitatively and selectively extracted with TOPO into toluene from 0.1 to 7 molar nitric acid medium which can be exploited both for its separation from complex matrices and for its sensitive estimation after colour reaction with 3,5 diBr PADAP in non-aqueous medium. So, a systematic investigation has been taken up on the extraction of Th(IV)-TOPO system to fix the optimum experimental conditions for its separation from various complex matrices and these results are presented in this section.

6.2 Experimental

6.2.1 *Reagents and Standards*

(1) Thorium (IV) Solution :

1×10^{-2}M Th(IV) solution is prepared by dissolving 1.380 gms of $Th(NO_3)_4\ H_2O$ (Fluka, Switzerland) in 50 ml of 7M HNO_3 and diluted to 250.0 ml. in a volumetric flask with 4M nitric acid. Thorium content is standardised complexometrically with EDTA. Working solutions are made by diiuting to suitable volume with distilled water.

(2) 3,5 diBr PADAP solutions is prepared as described in 3.2

(3) TOPO Solution :
Tri-n-octyl phosphine oxide (TOPO) (Fluke, Switzerland) : 0.05M (TOPO) solution is prepared by dissolving 7.733 gms. in 250 ml. carbon tetrachloride.

(4) Sodium Nitrate :
0.68 gm. of sodium nitrate is dissolved in one litre of double distilled water to get 8 molar sodium nitrate.

(5) Toluene : Double distilled toluene (BDH) is used.

All other chemicals used are of analar grade.

The optimum conditions for the quantitative extraction of thorium are established by following extraction of Th(IV) with tri-n-octyl phosphine oxide under systematically varied experimental conditions by adopting general extraction procedure given below :

6.2.2 *General Extraction Procedure*

Transfer X.0 ml of 1 X 10^{-3}M thorium (IV) solution in a separatory funnel, add 5.0 ml of 8 M $NaNO_3$ and x ml of concentrated nitric acid to give required overall acidity. Make-up the aqueous layer to 25.0 ml with distilled water. Equilibrate this solution for ten minutes with 25.0 ml of 0.05 M TOPO in toluene. After clear separation of the two phases, remove the organic extract and transfer an aliquot of the organic extract to a 10 ml volumetric flask and add 0.5 ml of 1 X 10^{-2} M 3,5 di Br PADAP in ethanol and 3 ml of pyridine, and keep this solution in standing for 15 minutes. Finally make-up the solution with absolute ethanol and shake well. Measure the absorbance of the resultant solution at 575 nm against similarly prepared reagent blank. Determine Th(IV) present in the solution by computing from standard Beer's law curve of Th(IV)-TOPO-3,5 di Br PADAP method given in 7.5.

Similarly estimate Th(IV) present in the aqueous phase also by Th(IV)-TOPO-3,5 di Br PADAP method.

6.2.3 *Selection of Solvent*

Many of the organic solvents such as hexane, chloroform, carbontetrachloride, cyclohexane, nitrobezene, benzene, methyl isobutyl ketone and toluene are tested for the extraction of thorium (IV). In these investigations Th(IV) is back extracted from the organic phase and then estimated by Th(IV)-TOPO-3,5 di Br PADAP method.

Of all these solvents toluene, carbontetrachloride, and cyclohexane are found to extract thorium quantitatively. However in the spectrophotometric studies on the colour reaction of Th(IV)-TOPO colourless complex with 3,5 di Br PADAP to yield intensely coloured complex Th (IV) - TOPO - 3,5 di Br PADAP complex in non aqueous medium, it is observed that the toluene medium is a facile medium for this reaction giving more intensely coloured species. So, all the extraction studies are carried out using tolune as solvent.

6.2.4 *Effect of Shaking Time on the Extraction of Th(IV)-TOPO Complex in Toluene*

It is observed that a 10.0 minute shaking period is required for attaining quantitative extraction of Th(IV) into toluene as Th(IV)-TOPO complex. If the shaking period is maintained at ten minutes the first extraction itself extracts 06% of thorium resulting in negligible extraction in the second extraction.

TABLE 6.1

Effect of Shaking Time on the Extraction of Th(IV)-TOPO-Complex into Toluene.

Aqueous Phase: 5:0 ml of 1 X 10^{-3} M Th(IV) solution + 5.0 ml of $NaNO_3$ (8 molar) + 10.5 ml of 12 M HNO_3 total volume made to 25.0 ml.

Organic Phase : 25.0 ml of 0.04 M TOPO.

S. No.	*Shaking time in minutes*	*%E of Th(IV)*
1	2	3
1.	1.0	14.3
2.	2.0	16.2
3.	3.0	18.6
4.	4.0	26.7
5.	5.0	33.1

(Contd.)

Table 6.1 (Contd.)

1	2	3
6.	6.0	67.8
7	7.0	84.2
8.	8.0	92.9
9.	9.0	95.6
10.	10.2	96.0
11.	11.0	96.0
12.	12.0	96.0
13.	15.0	96.0

6.2.5 *Effect of Acidity*

The effect of acidity of the medium on the extraction of thorium as Th(IV)-TOPO complex is studied by following the extraction at a fixed concentration of Th(IV) 5 ml 1 X 10^{-3} with large excess of TOPO (5 X 10^{-2}M) by varying acids and also their concentrations.

TABLE 6.2
Effect of Acid Concentration on the Extraction of Th(IV)-with TOPO into toluene

Aqueous Phase : 5.0 ml of 1 X 10^{-3} M Th(IV) + 5.0 ml of 8M $NaNO_{3+}$ + X.0 ml of conc. HNO_3 or conc. HCl or conc. H_2SO_4 or conc. $HClO_4$ total volume : 25.0 ml.

Organic Phase : 25.0 ml of toluene (0.05 M TOPO)

S. No.	*Molarity of acid*	*% Extraction of Th*			
		HCl	*HNO_3*	*H_2SO_4*	*$HClO_4$*
1	2	3	4	5	6
1.	0.1	11.2	74.7	18.2	16.7
2.	0.2	19.4	76.4	21.6	18.3
3.	0.3	26.8	78.0	27.2	23.7
4.	0.4	32.5	80.2	33.7	26.2
5.	0.5	37.1	85.4	36.8	30.4
6.	0.6	48.9	89.2	44.2	33.2
7.	0.7	54.6	93.6	56.1	38.2

(Contd.)

Table 6.2 (Contd.)

1	2	3	4	5	6
8.	0.8	59.8	99.9	68.2	41.3
9.	0.9	65.1	99.9	74.2	44.6
10.	1.0	71.7	99.9	79.8	52.8
11.	2.0	79.4	99.9	82.1	61.6
12.	3.0	87.3	96.0	77.3	62.8
13.	4.8	94.6	96.0	62.1	63.4
14.	5.0	99.9	96.0	56.7	58.6
15.	6.0	99.9	96.0	53.3	49.2
16.	7.0	99.9	96.0	41.4	33.1

The results indicate that thorium (IV) can be quantitatively extracted from 2-3 M HNO_3 and 5-7 M HCI with TOPO 0.05 M. However, in nitric acid medium even upto 7 M, though the extraction of thorium slightly decreases it is fairly constant and never lower than 90%. With H_2SO_4 or $HCIO_4$ the extraction of thorium is not quantitative and the results are found not be reproducible particularly in high acid concentration.

To examine the best acidity range with respect to nitric acid for the selective extraction of Th(IV), with TOPO the extraction of other likely associated metal ions like Cu(II), Zn(II), Bi(III), U(VI), Zr(IV), Mo(VI), Fe(III), V(V) and As(III) with TOPO are examined in 1-7 M HNO_3 medium. Percent extraction of different metal ions at varying acidities are plotted in Fig. 6.1 along with that of thorium is quantitative with TOPO from 2-3 M nitric acid medium. However, the extraction is highly selective and free from interferences at 4-7 M acidity range though it is slightly less (95%). So, to make the thorium extraction highly selective 5 molar nitric acid medium has been chosen and all further investigations for analytical studies on thorium are carried out at 5 M nitric acid medium only.

6.2.6 *Effect of Salting-Out Agents and Effect of Dilution*

The effect of salting-out agents using electrolytes like sodium nitrate, lithium nitrate, aluminium nitrate, chloride, sulphate on the extraction of thorium at constant acidity with 0.04 M TOPO has been studied Fig. 6.2. These results indicate that the addition of nitrate or chloride ions at low acidity give rise to high distribution coefficients probably because of salting out action. However, at higher acidities they

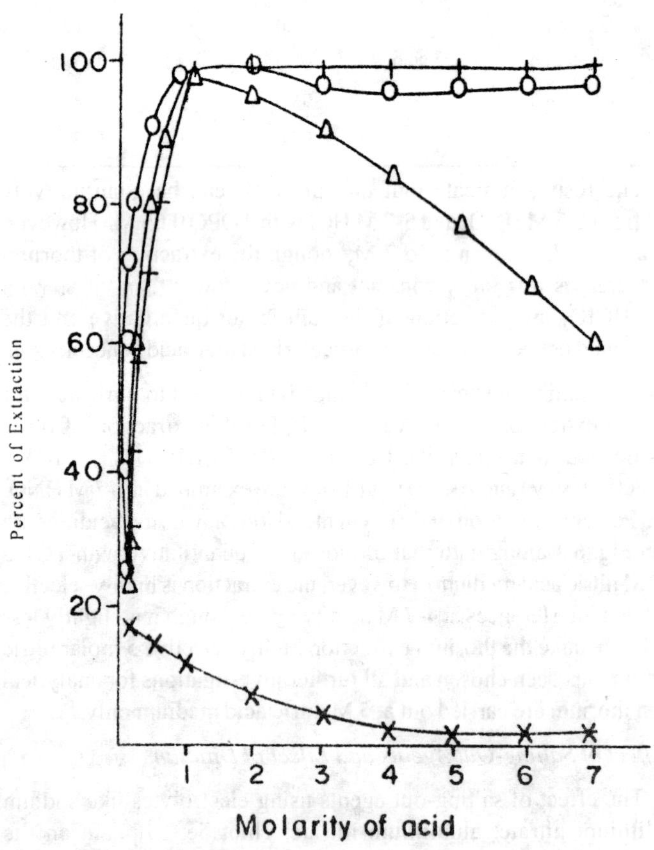

Fig. 6.1 : Extraction of metal ions from nitric acid medium by 0.05M *TOPO*.

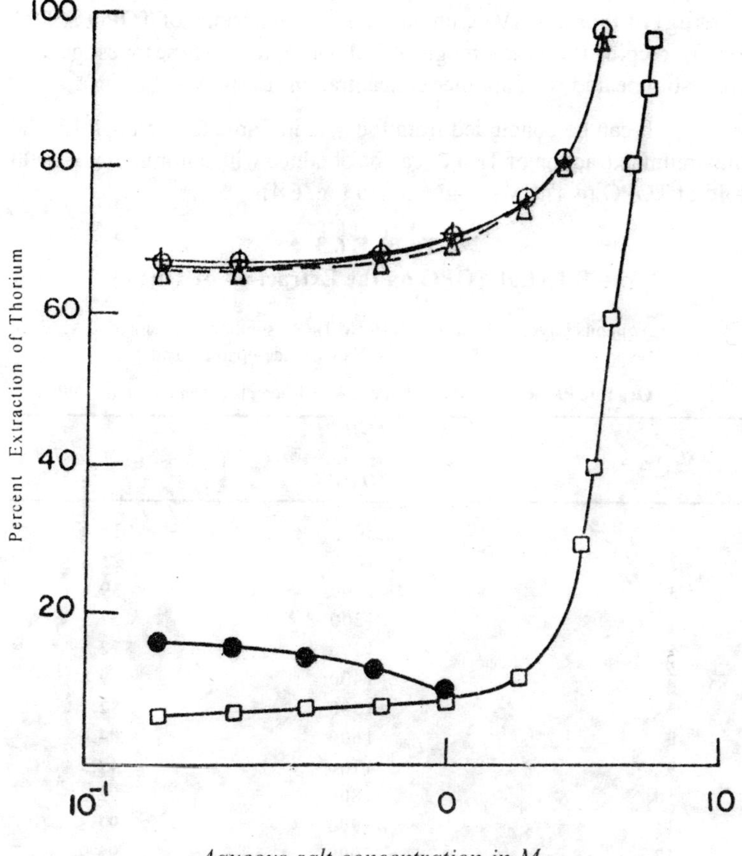

Fig. 6.2 : Effect of salting out agents on the extraction of thorium by 0.04M TOPO toluene Th(iv) – $3x\ 10^{-3}$ M

have little effect on the extraction. Addition of sulphate ions gives rise to steady decrease in the distribution. There is no effect of dilution on the quantitative extraction of Th even when aqueous to organic volumes are varied from 1:1 to 10:1.

6.2.7 *Effect of Concentration of TOPO on the Extraction of Th(IV)-TOPO in Toluene.*

The effect of TOPO concentration on the extraction of Th(IV) (750 μg) into toluene is investigated by extracting a fixed amount of (750 μg) of thorium (IV) with varying concentrations of TOPO at 5 M HNO_3, keeping the ionic strength at 2 M with $NaNo_3$. These investigations are also repeated with another concentration of Th(IV) (1.6 mg).

It can be concluded from the data in Table 6.3 and 6.4 that the maximum extraction of Th(IV) can be obtained with a minimum of 2000 fold of TOPO to Th(IV) (Fig. 6.3 and Fig. 6.4).

TABLE 6.3
The Effect of TOPO on the Extraction of Th(IV)

Aqueous Layer : 5.0 ml of 1 X 10^{-3} M Th(IV) solution + 10.5 ml of 12 M HNO_3 + 5.0 ml of 8 M $NaNo_3$ made-up to 25.0 ml.

Organic Phase : 25.0 ml of toluene with different concentrations of TOPO.

Sl. No.	*[TOPO] / [Th(IV)]*	*% E of Th (IV)*
1.	500	11.3
2.	1000	21.3
3.	1100	39.8
4.	1200	57.3
5.	1300	72.8
6.	1400	79.8
7.	1500	82.1
8.	1600	84.4
9.	1700	47.3
10.	1800	92.2
11.	1900	94.3
12.	2000	95.9
13.	2100	96.2
14.	2200	96.1
15.	2500	96.3

TABLE 6.4
The Effect of TOPO on the Extraction of Th(IV) (1.6 mg)

Aqueous Layer : 7.0 ml of 1 X 10-3 M Th(IV) solution + 10.5 ml of 12 M HNO_3 + 5.0 ml of 8 molar sodium nitrate, made upto 25.0 ml.

Organic Phase : 25.0 ml of toluene with different concentrations of TOPO.

Sl. No.	*[TOPO] / [Th(IV)]*	*% E of Th (IV)*
1.	700	14.6
2.	800	19.1
3.	900	39.1
4.	1000	66.2
5.	1100	78.5
6.	1200	80.2
7.	1300	84.4
8.	1400	86.7
9.	1500	88.1
10.	1600	90.4
11.	1700	92.1
12.	1800	92.3
13.	1900	94.6
14.	2000	95.8
15.	2100	96.2
16.	2200	96.4
17.	2500	96.4

6.3 Interference Studies

The effect of various likely associated ions on the extraction of Th(IV)-TOPO complex is studied by extracting 163 μg of Th(IV) with TOPO at 5 M nitric acid medium into toluene in presence of various foreign ions by adopting the general extraction procedure. In HCI medium also similar studies are made. The corresponding results are given in Table 6.5.

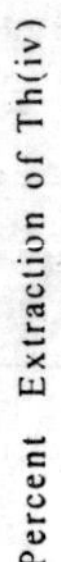

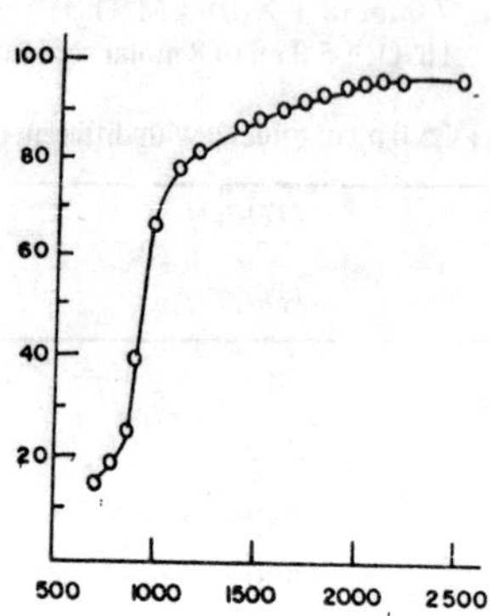

Fig. 6.3 : Effect of varying molar excess of TOPO on the extraction of Th(iv) from 5M nitric acid medium.

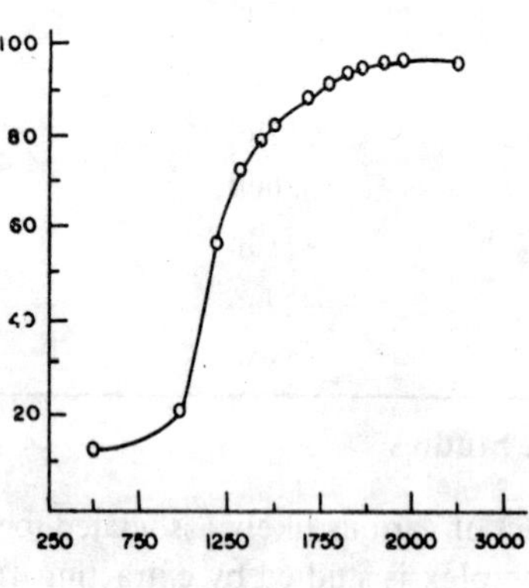

Fig. 6.4 : Effect of varying molar excess of TOPO on the extraction of Th(iv) from 5M nitric acid medium.

TABLE 6.5

Effect of Various Diverse Ions on the Extraction of Th(IV)-TOPO Complex into Toluene.

Aqueous Phase : one ml of 1 X 10^{-3} M Th(IV) solution + 5.0 ml of 8 M $NaNO_3$ + X.0 ml of conc. HNO_3 over all acidity 5 molar, total volume made-up to 25.0 ml.

Organic Phase : 25.0 ml of 0.05 M TOPO in toluene.

S. No.	*Foreign ion/ compound*	*Source material*	*Amount added*	*Percent extraction of Th*	
				HNO_3	*HCl*
1	2	3	4	5	6
1.	–	–	–	96.3	98.3
2.	F^-	NaF	15 mg	96.0	98.8
3.	Cl^-	KCl	100 mg	96.2	98.2
4.	NO_3^-	$NaNO_3$	100 mg	96.0	90.3
5.	PO_4^{-3}	H_3PO_4	100 mg	95.8	98.6
6.	SO_4^{2-}	Na_2SO_4	10 mg	96.4	98.6
7.	Oxalate	$(COONH_4)_2H_2O$	50 mg	95.9	98.2
8.	Tartrate	$C_4H_4O_6Na_2$	50 mg	96.4	98.6
9.	Nitrite	$NaNO_2$	10 mg	96.3	98.1
10.	Hydrogen peroxide	H_2O_2	one ml 38 volumes	Interfered	Interfered
11.	Ascorbic Acid	$C_6H_{70}6$	10 mg	96.3	98.1
12.	Diethyl dithio carbamate	$(C_2H_5)2$ $NCS_2Na.\ 3H_2O$	100 mg	95.9	Interfered
13.	EDTA	$C_{10}H_{18}O_{10}$ N_2Na_2	0.5 mg	Interfered	Interfered
14.	Fe (III)	$FeCl_3$	100 mg	96.3	Interfered
15.	Co (II)	$Co\ (NO_3)_2$	3 mg	95.9	98.7
16.	Cr (VI)	$K_2Cr_2O_7$	1 mg	Interfered*	Interfered*
17.	Al (III)	$Al\ (NO_3)_2$	7 mg	95.9	98.7
18.	Zr (IV)	$Zr\ (NO_3)_4$ $5\ H_2O$	40 µg 45 µg	95.2 Interfered*	98.7 Interfered*
19.	U (VI)	$UO_2(NO_3)_2$	45 µg	Interfered*	Interfered*
20.	Nb (V)	Nb_2O_5	25 µg	Interfered	Interfered

(Contd.)

Table 6.5 (Contd.)

1	2	3	4	5	6
21.	Ti (IV)	Titanium Sponge	75 µg 80 µg	95.2 Interfered*	Interfered* Interfered*
22.	La (III)	$LaCl_3 7H_2O$	200 µg 250 µg	96.1 Interfered*	Interfered* Interfered*
23.	Ce (IV)	$CeSO_4$	200 µg 250 µg	96.2 Interfered*	Interfered* Interfered*
24.	Y (III)	Y_2O_3	100 µg 145 µg	95.7 Interfered*	Interfered* Interfered*
25.	Zn (II)	$Zn(NO_3)_2$	3 mg	95.9	Interfered*
26.	V (V)	NH_4VO_3	2 mg	95.9	Interfered*
27.	Sn (IV)	Na_2SnCl_6	10 µg	95.7	Interfered*
28.	Mo (VI)	$(NH_4)_2 MoO_6$	1 mg 10 mg	96.3 95.7	Interfered* Interfered*
29.	Cu (II)	$Cu(NO_3)_2$ $6 H_2O$	3 mg	95.9	Interfered*
30.	Ca (II)	$CaCl_2 6H_2O$	4 mg	96.1	98.7
31.	Mg (II)	$MgCl_2 6H_2O$	18 mg	95.8	98.3
32.	Bi (III)	$Bi(NO_3)_3$ $5 H_2O$	4 mg	96.3	Interfered*
33.	Pb (II)	$Pb(NO_3)_4$	3 mg	95.8	Interfered*
34.	As (III)	As_2O_3	4 mg	95.9	Interfered*
35.	Mn (II)	$MnCl_2$	3 mg	95.7	Interfered*
36.	Ni (II)	$Ni(NO_3)_2$ $6 H_2O$	4 mg	96.3	98.7
37.	Mn (VII)	$KMnO_4$	1 mg	Interfered*	Interfered*

* *Positive Error*

These investigations revealed that chloride (100 mg), nitrite (10 mg), nitrate (100 mg), tartrate (50 mg), oxalate (50 mg), ascorbic acid (10 mg), fluoride (15 mg), Co(II) (3 mg), Al(III) (7 mg) Ca(II) (4 mg) and Mg(II) (18 mg) do not interfere in the extraction of Th(IV) with TOPO both in nitric acid or hydrochloric acid.

Similarly hydrogen peroxide (one ml of 38 volumes), EDTA (0.5 mg) U(VI) (45 µg) Zr(IV) (45 µg) Nb(V) (25 µg) interfere both in nitric acid and hydrochloric acid medium.

However, in nitric acid medium effective extraction of Th(IV) with TOPO can be achieved even in presence of large concentrations of diethyl dithiocarbamate (100 mg), Fe(II) (100 mg), Cu(II) (3 mg). Zn(I) (3 mg), Mo(VI) (10 mg), As(III) (4 mg), Mn(II) (3 mg), V(V) (2 mg), and Bi(III) (4 mg). While smaller concentrations of Ti(IV) (75 μg), Sn(IV) (10 μg) La(III) (200 μg), Ce(IV) (200 μg), Y(III) (100 μg) can be tolerated. When the extraction is carried out in hydrochloric acid medium the above ions seriously interfere in the extraction of Th(IV) with TOPO.

In general it is observed that the extraction of Th(IV) with TOPO from 5 M nitric acid medium is free from interference of more diverse substances and the tolerance limits of certain other substances are also found to be high when compared to the extraction in HCl medium. The interesting observation in these investigations is that complexing anions like fluoride, diethyl dithiocarbamate, tartrate and oxalate which seriously interfere in many other methods for the extraction of thorium and which can be used as masking agents for several metal ions can be tolerated even at high concentration.

6.4 Studies on the Nature of Th-TOPO Extracting Species

The extraction of $Th(NO_3)_4$ with different concentrations of TOPO is followed at 1 M HNO_3 keeping the overall ionic strength of $NaNO_3$ at 2 M and at temperature 25°C. In view of the ease of extraction of nitric acid with Th-TOPO and the relatively high acidity condition the concurrent, co-extraction of acid along with Th-TOPO occurs. The extraction of free nitric acid with TOPO can be described by the following generalized equation.

$$H^{+} + NO_3^{-} + n\,(TOPO)_o \rightleftharpoons [HNO_3 n(TOPO)]_o \quad \ldots\ldots \quad (1)$$

The simplest generalized equation for the extraction of thorium nitrate with TOPO can be written as follows :

$$Th^{4+}(a) = 4\,NO_3^{-} + n\,TOPO \rightleftharpoons Th(NO_3)_4\, n\,TOPO \quad \ldots\ldots \quad (2)$$

If all concentrations are small relative to nitrate, i.e., the nitrate concentration is relatively unchanged before and after equilibrium n can be determined by a plot of log D_a Vs. log (TOPO) where D_a, the extraction coefficient, is defined as follows :

$$D^{o}_{a} = \frac{\text{Thorium concentration in the organic phase}}{\text{Thorium concentration in the aqueous phase}}$$

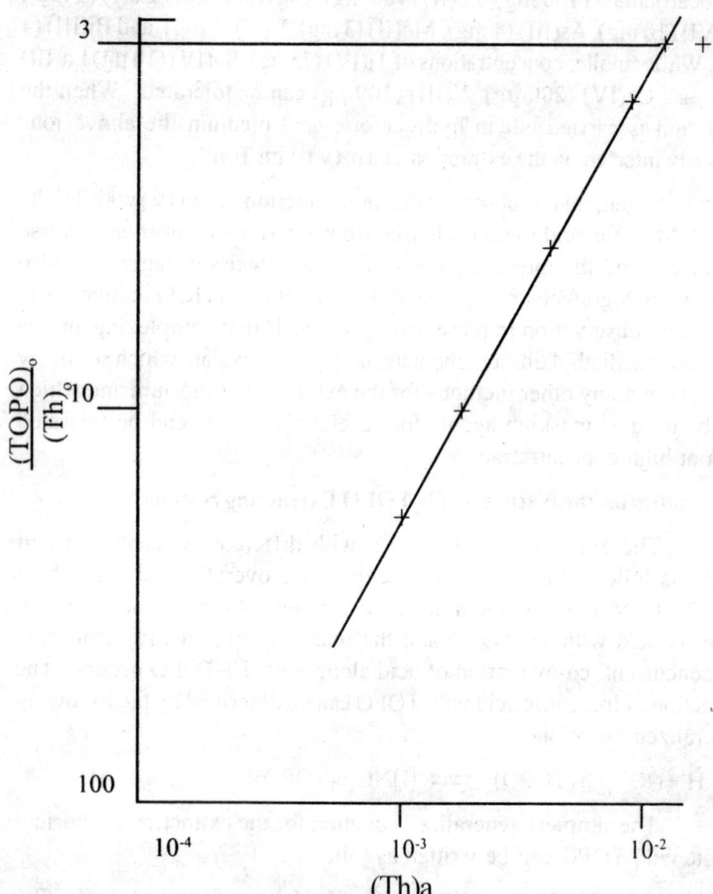

Fig. 6.5 : Ratio of TOPO to thorium concentration in toluene as a function of equilibrium thorium concentration.

and (TOPO) represents the free, uncomplexed, equilibrium concentration of the organic reagent.

In Fig. 6.5 are shown the results obtained in the present investigation when the equilibrium aqueous thorium concentration plotted against the equilibrium ratio $(TOPO)_o / (Th)_o$. The value of $(TOPO)_o$ represents the initial organic reagent concentration minus the nitric acid concentration of the organic phase. The latter correction is made with the assumption that one mole of TOPO is associated with each mole of nitric acid extracted, at low acidities and high ionic strength with sodium nitrate. The thorium ion concentration was varied from 0.0002 to 0.5M while maintaining the initial acidity at 0.1M. The latter value was as low as 0.07M at equilibrium because the amount of acid extracted increased with decreasing concentration of thorium. At higher thorium concentrations 0.03 - 0.05 M the organic reagent is essentially completely combined with the acid and metal salt so that calculations of the free, equilibrium concentration of TOPO were not made. At lower concentration of thorium 0.002 M, the equilibrium concentration of thorium in the aqueous phase was low, i.e., 10^{-6} M, and the value of extraction co-efficient and the equilibrium quotients are subjected to considerable analytical error.

A test of the dependence of the extraction on the TOPO concentration was made by varying the initial reagent concentration from 0.1 to 0.00025 M at constant initial aq. conc. of Th and (H^+) ion. The acidity was maintained at lower level, so as to minimize complications resulting from extraction of acid. Assuming the formation of a 3:1 TOPO: Th complex from equation no. 2) the following relationship is obtained.

$$\text{Log } K = \log D°a - 4 \log (NO^-_3) + 3 \log (TOPO) \quad \ldots\ldots \quad (3)$$

If the formation of a 3:1 complex is correct, values of K calculated on the basis of this assumption should be constant. Also a plot of log D°a against log (TOPO) at constant nitrate ion concentration should have a slope of three.

As direct measure of the free TOPO, or of the organic hydrogen ion concentration, was not available both these values are calculated by difference. The equilibrium conc. of free TOPO $(TOPO)_e$ was calculated as follows :

$$(TOPO)_e = (TOPO)_i - 3(Th)_o + (H^+)_o \quad \ldots\ldots \quad (4)$$

Where $(TOPO)_i$ is the initial concentration of TOPO and $(Th)_o$ and $(H^+)_o$ are equilibrium concentrations of Th(IV) and H^+ ion in the organic phase. The errors introduced by the use of values based on differences will be large in those cases where the TOPO is essentially all complexed, but should be small in the more concentrated reagents solutions.

Assuming the formation of the 3:1 complex and a constant nitrate ion concentration of 2M the following simplified form

$$K = \frac{D^{\circ}a}{16\,(TOPO)^3} \quad \ldots\ldots \quad (5)$$

K is calculated using the data in the Table 6.6 Values of K were not calculated in those cases where D°a was too indefinite due to the extremely dilute aqueous thorium levels at high organic concentrations. In addition K values could not be calculated when the analysis showed that the concentrations of the extracted species equalled or exceeded the reagent concentration, where the data permitted for the calculation of K values, the general agreement is quite satisfactory.

At higher acidities, the extraction of acid by the organic reagent is appreciable and the thorium extraction is no longer independent of the acid extraction. The TOPO. HNO_3 complex may itself act as the extracting agent leading to adducts of the type $[Th(NO_3)_4 3\ TOPO]n\ HNO_3$ TOPO].

Thus the formation of a Th $(NO_3)_4$ 3 TOPO complex seems to be fairly well established provided that the aqueous phase is fairly dilute with respect to both thorium and acidity. At higher concentrations and acidities other species like $[Th(NO_3)_4\ 3\ TOPO]n\ HNO_3$ TOPO may be responsible for extraction of Th(IV).

Table - 6.6
Thorium Nitrate Extraction as a Function of Variable TOPO Concentration

S. No.	$(Th)_a$	$(Th)_o$	D°_a	$(H^+)_a$	$(H^+)_o$	$(TOPO)_i$ *M*	$(TOPO)_e$	*K*
1.	0.0058	0.0036	0.6200	0.0092	0.0003	0.0127	0.00160	9×10^6
2.	0.0069	0.0024	0.3500	0.0092	0.00050	0.0092	0.00150	6×10^6
3.	0.0082	0.0016	0.1900	0.0095	0.0004	0.0065	0.00140	4×10^6
4.	0.0090	0.0008	0.0800	0.0095	0.0002	0.0035	0.00090	7×10^6
5.	0.0095	0.0009	0.0094	0.0097	0.0001	0.0008	0.00043	7×10^6

7

Spectrophotometric Studies on Th (IV)-TOPO-3,5-diBr PADAP Ternary System

7.1 Introduction

Because of the complexity of the composition of ores, rocks and minerals containing thorium, the determination of thorium almost always involves isolating it from accompanying elements. Depending on the nature of the problem, thorium may be determined by gravimetric, volumetric or colourimetric methods. Gravimetric methods and volumetric methods are most often recommended when the highest accuracy is not required. For the determination of very small quantities, spectrophotometric methods are used. Many of the available methods lack sensitivity and selectivity.

The various chromogenic reagents used to determine thorium are chromotrope 2B[77] eriochrome Black T[78, 79] solochrome fast Red[80], xylenol orange[81-83], methylthymol blue[84, 85], chromozurol S[86, 87] and quercetin[88]. Some of these reagents suffer from less sensitivity and some from the lack of selectivity when applied for the analysis of complex materials.

In the author's investigation, it is observed that the selectively extracted Th(IV)-TOPO colourless species from 5M nitric acid medium can be converted into an intensely coloured ternary species by reacting with a chromogenic complexing agent like 4 (2-Pyridylazo) resorcinol (PAR), 1 (2-pyridylazo) - 2 -naphthol (PAN), PADAP, 5-Br PADAP and 3,5 di Br PADAP in nonaqueous medium. As this colour reaction with 3,5 di Br PADAP is more sensitive and selective than with other chromogenic complexing agents, this colour reaction is systematically studied to exploit it for trace spectrophotometric determination of thorium in complex materials like rocks, minerals, ores and alloys and the results are presented in this chapter.

EXPERIMENTAL

7.2 General Extraction and Colour Development Procedure

Transfer 5.0 ml of 6.4 x 10^{-5} M Th(IV) solution into a separatory funnel and add 8.5 ml of 12 molar nitric acid and make-up the aqueous layer to 20.0 ml. Equilibrate for ten minutes with 20.0 ml of 0.04 M TOPO in toluene. After clear separation of the two phases, remove the organic extract and transfer an aliquot (1-4 ml) of this to a 10.0 ml volumetric flask, and 0.5 ml of 1 x 10^{-2} M 3,5 di Br PADAP in ethanol and 3.0 ml of pyridine. Allow this solution to stand for colour development for 15.0 minutes. Make-up the volume to 10.0 ml with absolute ethanol. Measure its absorbance at 575 nm against similarly prepared reagent blank.

7.2.1 *Selection of Best Chromogenic Complexing Reagent for Ternary Complex Formation*

It is observed that Th(IV)-TOPO complex extracted into toluene can be converted into a mixed ligand complex of high colour sensitivity by adding a second chromogenic complexing agent like 4 (2-pyridylazo) resorcinol (PAR), 1 (2-pyridylazo)-2naphthol (PAN), PADAP, 5-Br PADAP and 3,5 di Br PADAP in pyridine + toluene + alcohol medium. Of these reagents, best results, (Table 7.1) with good reproducibility and high sensitivity are observed with 3,5 di Br PADAP which is used in all further investigations. Further addition of pyridine to the medium accelerates the colour formation and is found to increase the sensitivity of the colour reaction.

7.2.2 *Effect of Base Concentration in the Medium on the Colour Formation of Th (IV)-TOPO-3,5 di Br PADAP Ternary Complex*

Solvent properties like dielectric constant, polarity, polarisability, directly affect the thermodynamics of complex formation reactions through solvent - solvent and solvent - solute interactions. An important chemical property of the solvent is the basicity of the solvent molecules. Compounds with the functional groups containing electronegative atoms such as oxygen or nitrogen increase the basicity with increase in electron density on these

TABLE 7.1

Selection of Best Chromogenic Complexing Reagent

Aqueous Phase: 5.0 ml of 6.4 x 10^{-5} M of Th(IV) solution + 8.5 ml of 12 M nitric acid, total volume 20.0 ml.

Organic Phase : 20.0 ml of 0 05 M TOPO in toluene. 4 ml of the organic extract + 1 ml of 3,5-diBr PADAP (ix 10^{-2}M)+3 ml of Pyridine + standing for 15 minutes and make up to 10 ml.

S. No.	*Chromogenic reagent*	*Absorbance max nm*
1.	1(2-Pyridylazo) 2 naphthol	0.531 (590)
2.	4(2-Pyridylazo) resorcinol	0.700 (555)
3.	PADAP	0.862 (560)
4.	5-Br PADAP	0.912 (570)
5.	3,5 di Br PADAP	1.20 (575)

atoms. It is observed that in the present investigations, presence of pyridine in the reaction medium accelerates the colour formation of Th(IV)-TOPO with 3,5 di Br PADAP and increases its sensitivity. So a systematic study has been taken up on the effect of various bases like pyridine, picoline, triethanalomine and aniline on the colour formation of Th(IV)-TOPO with 3,5 di Br PADAP. Of the various bases tested, pyridine gave best results by accelerating the reaction and increasing the colour sensitivity. Further, when pyridine composition is varied from 10 to 50% in the total non-aqueous medium. It is observed that 30 - 50% pyridine in the medium gives maximum absorbance with minimum time and further increases even up to 50% has no effect on the absorbance of Th(IV)-TOPO-3,5 di Br PADAP complex. It has been further observed that to the toluene extract taken in 10.0 ml volumetric flask first 3,5 di Br PADAP in ethanol solution is to be added, followed by addition of 3 ml of pyridine. The whole mixture is to be kept standing for full colour development for about fifteen minutes and then the mixture has to be diluted upto the mark with ethanol. This order of addition of reagents has to be strictly followed for achieving reproducible optical density values with maximum sensitivity for thorium. To get the maximum absorbance, it has been observed that a period of 15 minutes standing time is required after the addition of pyridine and before the addition of the absolute ethanol. Increase in standing time above 15 minutes did not produce any significant change in the absorbance even upto 1 hour.

Among various diluents tested such as acetone, absolute ethanol, amyl alcohol, 1,4 dioxane and toluene, absolute ethanol gave maximum absorbance values.

7.2.3 *Effects of Traces of Water in Ethanol*

No adverse effect of traces of water in ethanol is observed on the absorbance of Th(IV)-TOPO-3,5 di Br PADAP complex, when ethanol is used as diluent.

As the presence of traces of water in ethanol does not affect the absorbance, there is no need of using absolute ethanol at every stage of formation of the Th(IV)-TOPO-3,5 di Br PADAP ternary system.

7.2.4 *Effect of Concentration of 3,5-diBr PADAP on the Absorbance of Th(IV)-TOPO-3,5 di Br PADAP Ternary System*

A 40 fold excess of 3,5 di Br PADAP in ethanol is sufficient for attaining maximum colour and thereafter addition upto 100 folds has no effect on the absorbance of Th(IV)-TOPO-3,5 di Br PADAP mixed ligand complex when measured against appropriate reagent blank. The results of the investigation on the effect of 3,5 di Br PADAP concentration on ternary system are given in Table 7.2. It has been observed that heating on water bath for even about ten minutes has no effect on colour formation.

TABLE 7.2

Effect of Concentration of 3,5-diBr PADAP on the Absorbance of Th(IV)-TOPO-3,5 di Br PADAP System

Aqueous Phase : 6.4 X 10^{-5} M (Th(IV) solution 5.0 ml 8.5 ml of 12 M nitric acid, total volume made-up to 20.0 ml.

Organic Phase : 20.0 ml of toluene containing TOPO of 0.05 M non-aqueous. medium : 4.0 ml of the organic extract + 2.0 of 3,5 di Br PADAP with different concentrations (10^{-4} - 10^{-3} M) + 3.0 ml of pyridine allowed to stand for colour development for 15.0 minutes total volume made-up to 10.0 ml with absolute ethanol.

S. No.	*3,5 di Br PADAP {Th(IV)}*	*Absorbance*
1	2	3
1.	5	0.19
2.	10	0.31
3.	15	0.42
4.	20	0.54

(Contd.)

Table 7.2 (Contd.)

1	2	3
5.	25	0.69
6.	30	0.77
7.	35	0.77
8.	40	0.77
9.	45	0.77
10.	50	0.77
11.	60	0.77
12.	70	0.77
13.	80	0.77
14.	100	0.77

7.3 Absorption Spectra

To understand the spectral characteristics of Th(IV)-TOPO-3,5 di Br PADAP ternary system, its absorption spectra are examined by recording its absorbance at different wave lengths under varied experimental conditions against corresponding TOPO-3,5 diBr PADAP reagent blank in tolune-pyridine-ethanol medium by adopting the general extraction procedure in 6.2.2. The spectra of the Th(IV)-TOPO-3,5-diBr PADAP system obtained at different acidities (of nitric acid) different 3,5 di Br PADAP concentrations and different pyridine compositions of the medium are recorded. Figs. 7.1, 2 and 3.

For comparison the absorption spectra of Th(IV)-3,5 di Br PADAP aqueous system against blank and 3,5 di Br PADAP reagent blank against solvent medium are also shown in Figure 7.4. It can be seen from the figure that the Th(IV)-3,5 di Br PADAP binary complex system shows maximum absorption at 520 nm, while Th(IV)-TOPO-3,5-diBr PADAP ternary system in non-aqueous medium (toluene-pyridine-alcohol) shows maximum at 575 nm which indicates bathochromic shift upon ternary complex formation. It is further observed that the reagent blank has no significant absorbance at 575 nm.

As the spectra of Th(IV)-TOPO-3,5 di Br PADAP ternary system obtained at different concentrations of 3,5 di Br PADAP in acid medium and pyridine concentrations exhibit maximum at 575 nm, it can be

	Acidity - [Th (iv)]
+	1.0 M - 1.5 x 10^{-4} M
□	2.0 M - 2.2 x 10^{-4} M
O	2.5 M - 3.2 x 10^{-4} M
Δ	3.0 M - 3.5 x 10^{-4} M
●	3.5 M - 4.2 x 10^{-4} M
▲	4.0 M - 5.2 x 10^{-4} M

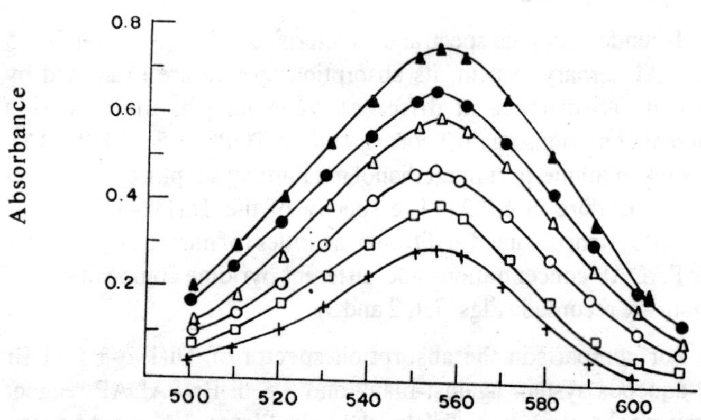

Wave length (nm)

Fig. 7.1 : Absorption spectra of Th(iv) - TOPO complex extracted into toluene at different acidities.

Th (IV) : 3,5 di Br PADAP

A + ----- + 1 : 8

B □ — □ 1 : 12

C O ----- O 1 : 15

D Δ — Δ 1 : 20

E O ----- O 1 : 25

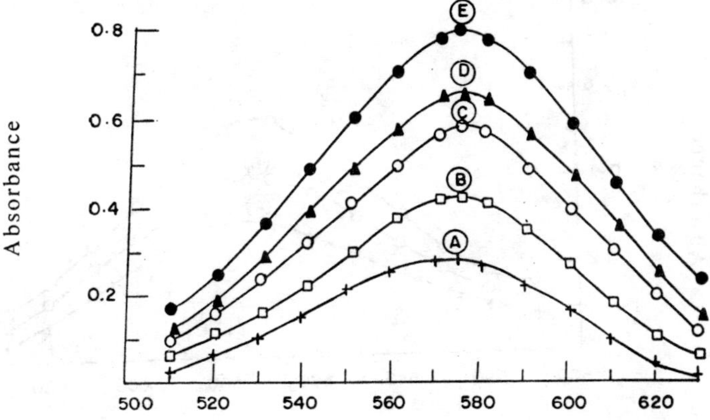

Wave length (nm)

Fig. 7.2 : Absorption spectra of Th(iv) - TOPO complex extracted into toluene from 5 molar nitric acid medium with different concentrations of 3,5 di Br PADAP

Th (IV) 6.0 ml of 1 x 10^{-4} M

Pyridine (in 10 ml)
A -- 10%
B -- 15%
C -- 20%

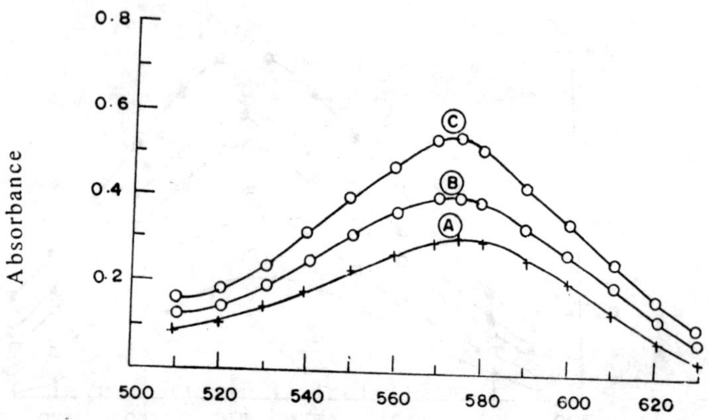

Wave length (nm)

Fig. 7.3 : Absorption spectra of Th(iv) - TOPO complex extracted into toluene from 5 molar nitric acid medium with different concentrations of pyridine.

[Th (IV) -4.0 x 10^{-4} M, 3,5 di Br PADAP 0.5 x 10^{-2} M]

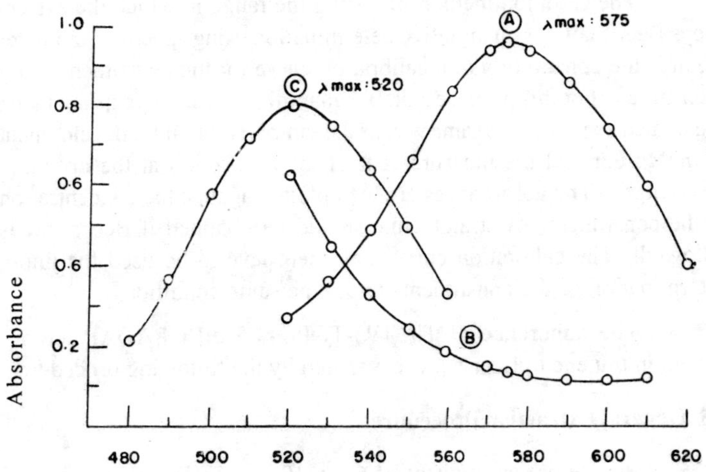

Wave length (nm)

Fig. 7.4 : Absorption spectra of

(A) Th(iv) - TOPO - 3,5 di Br PADAP mixed ligand complex 575 against reagent blank (Toluene, ethanol, pyridine medium) [Th(IV) = 6.5 ml of 1 x 10^{-4} M, 3,5 di Br PADAP 1x10^{-2}M. TOPO 0.05 M]

(B) TOPO - 3,5 di Br PADAP complex against toluene pyridine ethanol medium.

(C) Th(IV) - 3,5 di Br PADAP complex in aqueous medium at 520 pH 6.7

[Th (IV) 1.5 x 10^{-3} M, 3,5 di Br PADAP 1 x 10^{-2} M] reasonably concluded that a single absorbing species exists over a wide range of experimental conditions.

7.4 Adherence to Beer's Law by Th(IV)-TOPO-3,5 di Br PADAP Ternary System, in Toluene

The common method of testing the range in which the system obeys Beer's law for quantitative determination using spectrophotometer requires the constitution of a calibration curve for the constituent being determined. For this purpose suitable quantities of constituents are taken and treated exactly in the same way as the sample solution for development of the colour and the measurement of the absorption at the optimum wavelength. The absorbances are then plotted against the concentration of the constituent. A straight line should be obtained if Beer's law is followed. The calibration curve may then have to be used for future determination of the constituents under the same conditions.

The adherence of Th (IV)-TOPO-3,5 diBr PADAP ternary system in toluene to Beer's law is verified by the following procedure :

7.5 General Extraction Procedure

Pipette out an aliquant of 6.4 x 10^{-5} M Th(IV) solution into a 125.0 ml separatory funnel. Add 11 ml of conc. nitric acid to maintain 5 molar acidity and make-up to 35.0 ml. Equilibrate the mixture for ten minutes with 20.0 ml of 0.05 M TOPO in toluene. After clear separation of the two phases, transfer the organic layer into a microburette and run 4.0 ml of this solution into a 10 ml voluetric flask, add 1.0 ml 1 x 10^{-2} M 3,5 di Br PADAP solution in absolute ethanol, 3.0 ml of pyridine and stand this solution for 15 minutes for colour development. Then make up this solution to 10 ml with absolute ethanol and measure the absorbance at 575 nm against similarly prepared reagent blank.

The data obtained by adopting the above procedure with different concentration of Th(IV) are given in Table 7.3 and graphically represented in Figure. 7.5

From the Figure 7.5, it can be seen that Th(IV)-TOPO-3,5 di Br PADAP ternary system obeys Beer's law upto 3.25 μg/ml.

7.6 Sensitivity of the Colour Reaction

The average molecular extinction co-efficient and specific absorptivity for Th(IV)-TOPO-3,5 di Br PADAP mixed ligand complex into toluene - ethanol medium are found to be 1.2×10^5 $1.mol.^{-1} cm^{-1}$ and 0.504 ml $g^{-1} cm^{-1}$.

TABLE 7.3

Aqueous Phase : X.0 ml of 6.4 x 10-5 M Th(IV) + 11 ml of conc. HNO_3 total volume : 35.0 ml.

Organic layer : 20.0 ml of 0.05 M TOPO in toluene

Non-aqueous medium : 4.0 ml of the above extract + 1.0 ml of 1 x 10-2 M 3,5 di Br PADAP + 3.0 ml of pyridine, stand this solution for 15 minutes and makeup to 10.0 ml with absolute ethanol.

S. No.	*Amount of Th(IV) in μg per 35 ml in initial aqueous phase*	*Amount of Th(IV) μg/ml in final organic phase*	*Absorbance 575 nm*
1.	14.85	0.30	0.15
2.	29.7	0.59	0.31
3.	44.55	0.89	0.46
4.	59.40	1.19	0.62
5.	74.25	1.49	0.77
6.	89.10	1.78	0.92
7.	103.95	2.08	1.08
8.	118.80	2.38	1.23
9.	133.65	2.67	1.39
10.	148.5	2.97	1.54
11.	163.35	3.27	1.69
12.	178.20	3.56	1.85
13.	193.05	3.86	1.95

The Sandells sensitivity of the colour reaction is found to be 0.00193 $\mu g/cm^2$

7.7 Optimum Range from Ayre's and Ringbom's Plot

The percent relative error (dc/c X 200) per one percent photometric error has been calculated and plotted against transmittancy (Ayre's plot). The way in which analysis error varies with the transmittance is shown graphically in Fig. 7.6. The corresponding data is given in Table. 7.4.

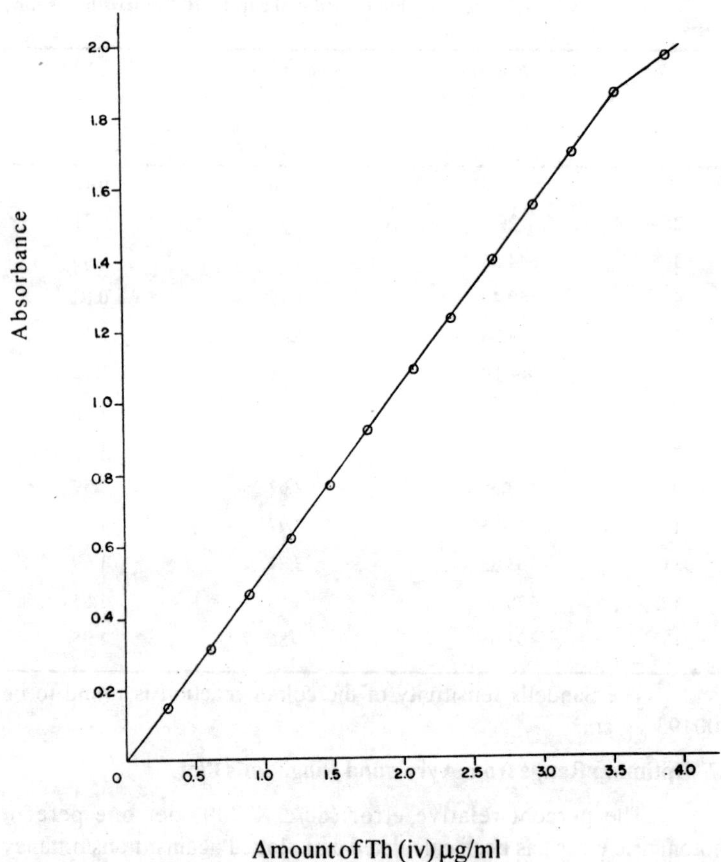

Fig. 7.5 : Adherence of Th(iv) - TOPO - 3,5 di Br PADAP method to Beer's law

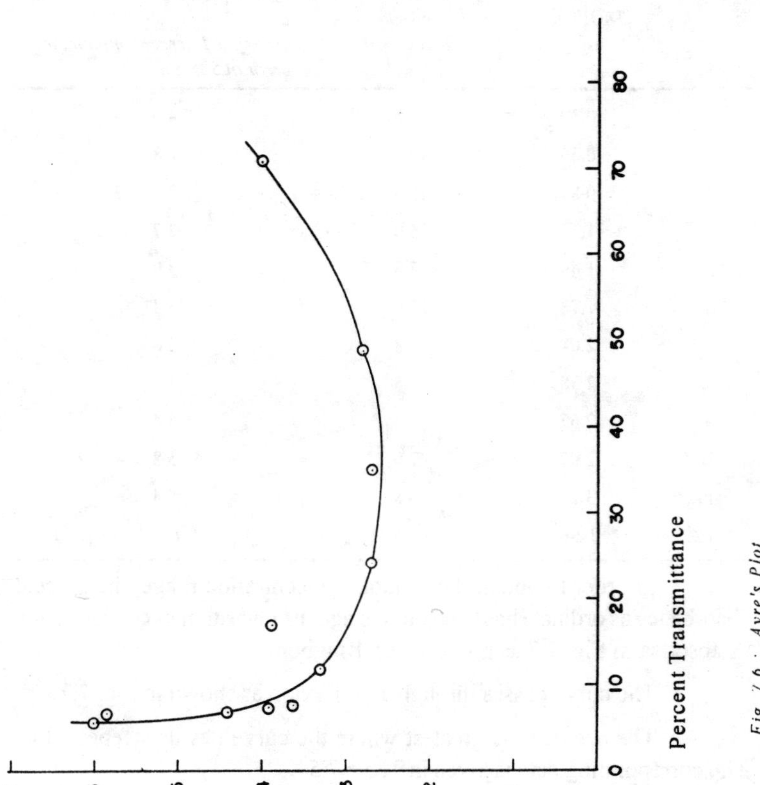

Fig. 7.6 : Ayre's Plot

TABLE 7.4
Data for Ayre's Plot

S. No.	*Amount Th(IV) in μg*	*Percent transmit-tance*	*$\frac{dc}{c}$ x 100 percent relative error per 1 percent photometric error at 575 nm*
1.	0.30	71.0	4.0
2.	0.59	49.0	2.8
3.	0.89	35.0	2.7
4.	1.19	24.0	2.7
5.	1.49	7.0	3.9
6.	1.78	12.0	3.3
7.	2.08	7.8	3.6
8.	2.38	7.5	3.9
9.	2.67	7.3	4.4
10.	2.97	7.0	5.8
11.	3.27	6.8	5.4
12.	3.56	6.5	6.0

In order to obtain the suitable concentration range, the percent absorbance as ordinate has been plotted against logarithm of concentration as abscissa in Fig. 7.7 as proposed by Ringbom's.

The curve passes through an inflexion as shown in Fig. 7.7

The accuracy is greatest where the curve has its steepest slop. The correponding data is given in Table 7.5.

$$di/dc/c = \frac{dI}{2.303} d \log c = \lambda \max$$

It is evident from the graphs (Ayre's and Ringbom's plots) that the optimum range with minimum photometric error is in the region 0.3 μg to 2 μg per ml of thorium with layer depth of one cm.

7.8 Precision Studies

The precision of the present method is evaluated by determining different aqueous concentrations of thorium (IV) in between 140 μg to 150 μg in 10 ml (each analysing six times) and the results are shown in Table

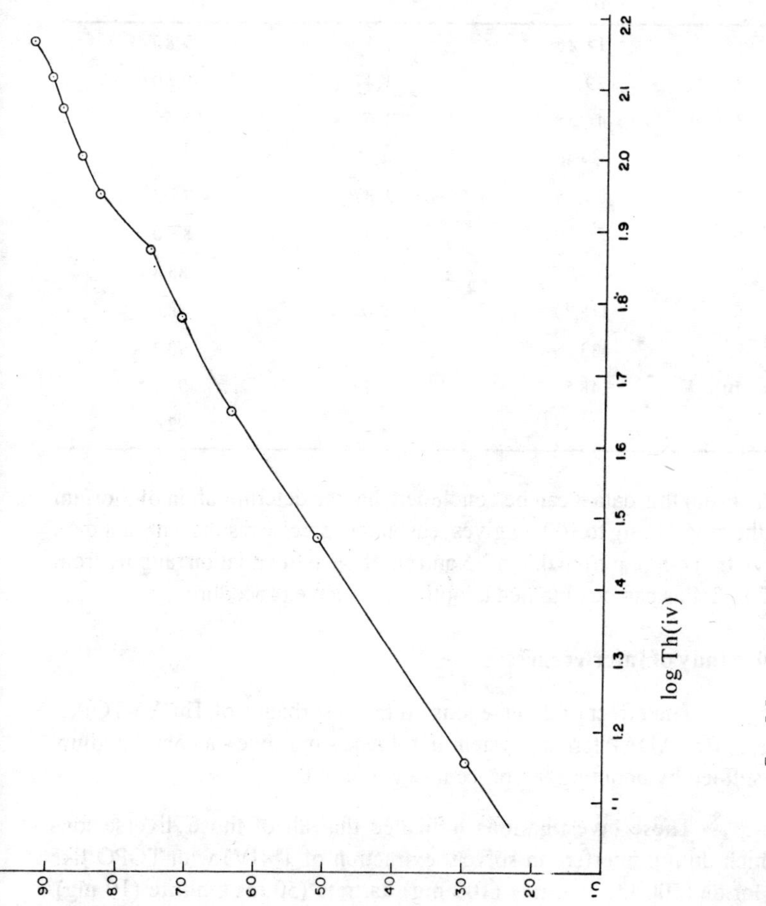

Fig. 7.7 : Ring bom's plot

TABLE 7.5
Data for Ringbom's Plot

S. No.	*Amount of Th(IV) μg*	*Log Th(IV)*	*Percent absorbance*
1.	14.85	1.17	29.86
2.	29.70	1.47	50.80
3.	44.55	1.65	63.57
4.	59.40	1.77	71.1
5.	74.25	1.87	75.7
6.	89.10	1.95	83.0
7.	103.95	2.02	85.8
8.	118.80	2.07	88.0
9.	133.65	2.12	90.2
10.	148.50	2.17	91.7
11.	163.35	2.21	92.4

7.6. From this data it can be concluded that the determination of thorium in the range 15 μg to 100 μg gives reasonably precise results with standard deviation ranging from 0.3 to 3.5 and coefficient of variation ranging from 0.2 to 2.4 as can be obtained adopting the above procedure.

7.9 Study of Interferences

The effect of diverse ions on the absorbance of Th(IV)-TOPO-3,5 di Br PADAP ternary system in toluene - pyridine - alcohol medium is studied by adopting the procedure given in 6.3.

These investigations indicated that all of those diverse ions which do not interfere in solvent extraction of Th(IV) with TOPO like chloride (100 mg.), nitrate (100 mg), tartrate (50 mg), nitrite (10 mg), oxalate (50 mg), ascorbic acid (10 mg), fluoride (15 mg), Co(II) (3 mg), Al(III) (7 mg), Cu(II) (3 mg), Mn(II) (3 mg), Ni(II) (4 mg). V(V) (2 mg), Ce(IV) (200 μg), Y(III) (100 μg), La(III) (200 μg), As(III) (4 mg), Bi(III) (2 mg), Mg(II) (18 mg), Ca(II) (4 mg) also do not interfere in Th(IV)-TOPO-3,5 di Br PADAP spectrophotometric method in non-aqueous medium. In general, the colour formation is more selective when it is extracted from nitric acid.

TABLE 7.6
Precision Studies on the Estimation of Th(IV)-TOPO-3,5-diBr PADAP Method

S. No.	*Conc. of Th (IV) in 20 ml in µg*	*(1)*	*(2)*	*(3)*	*(4)*	*(5)*	*(6)*	*Mean µg*	*S.D.*	*RSD = SD/Mean X 100*	*Mean ± found*
1.	23	22.4	23.4	23.3	23.2	23.5	23.2	23.16	0.4	1.7	23.16 ± 0
2.	46	46.5	46.6	46.8	45.8	46.7	46.5	46.48	0.4	0.8	46.48 ± 0
3.	63	63.9	62.1	62.2	63.6	62.3	63.4	62.90	0.8	1.2	62.90 ± 0
4.	92	90.6	89.9	92.8	94.6	95.5	93.1	92.75	2.2	2.4	92.75 ± 2
5.	115	115.7	112.6	118.8	113.6	119.1	115.3	115.85	2.7	2.3	115.85 ± 2
6.	138	138.7	138.6	138.9	139.1	138.3	139.1	138.75	0.3	0.2	158.75 ± 0
7.	161	156.6	165.4	158.1	164.3	162.1	163.8	161.70	3.4	2.1	161.70 ± 3

Average of six determinations. Confidence limit = Meant ± t S/N where S is the Standard Deviation, N are the replicate measurements and t is the statistical factor that depends on the number of degrees of freedom at 95% confidence level and for five degrees of freedom (N-1), t = 2,571.

It has been observed in the author's investigation that the diethyl dithiocarbamate which forms quantitatively extractable colourless chelates does not react with thorium and Th(IV)-TOPO-3,5 di Br PADAP method can tolerate large concentration of diethyl dithiocarbamate (100 mg). This situation has been exploited for the bulk separation of matrix and constituent components likely to be associated with thorium.

Further, other seriously interfering ions like U(VI), Ti(IV) and Nb(V) also can be tolerated in equal times to Th(IV) by preextracting them with DDTC into $CHCl_3$: $(CH_3)_2Co$ mixture at pH 1.0. However, Zr(IV) interferes in this method even in equal times to Th.

7.10 Jobs Method of Continuous Variation for the Determination of Th(IV) to 3,5 diBr PADAP in Th(IV)-TOPO-3,5 di Br PADAP System

The composition of Th(IV) to 3,5 di Br PADAP in the Th(IV)-TOPO-3,5 di Br PADAP system is determined by examining the variation in the absorbance with change in Th(IV)-3,5 di Br PADAP ratio by Job's method of continuous variations following the procedure described below :

To 5.0 ml of 1 x 10^{-2} M Th(IV) solution placed into a 125 ml separatory funnel, 11.0 ml of concentrated nitric acid is added and made upto 35.0 ml with distilled water. It is then equilibrated with 20.0 ml of 0.05 M TOPO in toluene for ten minutes. After clear separation of the Two phases, the toluene layer is removed and transferred into a microburette. 10.0 ml of this is placed into 50 ml volumetric flask and made up to the mark with toluene. X.0 ml of this solution is placed into a 25.0 ml. volumetric flask. To this (10 - X) ml of 0.0005M 3,5 di Br PADAP in absolute ethanol is added. 3.0 ml of pyridine is then added and this solution is kept standing for about 15 minutes and made up to the mark with absolute ethanol. It's absorbance is then measured at 575 nm against similarly prepared reagent blank. These results are given in Table 7.7 and Figure. 7.8.

The similar process is repeated with another concentration of thorium and 3,5 diBr PADAP (0.0006M). These results are also graphically represented in Figure 7.9 and the results are given in Table. 7.8.

From the nature of these Job's curves, which show maximum absorbance at 0.33 mole fraction in the final coloured species the ratio of Th:3,5-diBr PADAP is 1:2.

TABLE 7.7

Determination of the Ratio of Th(IV)-3,5 di Br PADAP Job's Method of Continuous Variations in Th(IV)-TOPO-3,5 di Br PADAP System

S. No.	*[Th (IV)]* / *[Th(IV)]} + 3,5 diBr PADAP]*	*Absorbance at 575 nm*	*Y scale = ABC / ABS .λ max at 575 nm*
1	2	3	4
1.	0.10	0.24	0.23
2.	0.20	0.43	0.42
3.	0.25	0.51	0.50
4.	0.30	0.58	0.57
5.	0.33	0.64	0.63
6.	0.40	0.62	0.61
7.	0.50	0.53	0.52

(Contd.)

Table 7.7 (Contd.)

1	2	3	4
8.	0.60	0.43	0.42
9.	0.70	0.28	0.27
10.	0.80	0.15	0.14
11.	0.90	0.06	0.05

TABLE 7.8
Determination of the Ratio of Th(IV)-3,5 di Br PADAP Job's Method of Continuous Variations in Th(IV)-TOPO-3,5 di Br PADAP System

S. No.	*[Th (IV)] / [Th(IV)} + 3,5 diBr PADAP]*	*Absorbance at 575 nm*	*Y scale = ABC / ABS .λ max at 575 nm*
1.	0.1	0.44	0.36
2.	0.2	0.59	0.48
3.	0.25	0.66	0.53
4.	0.3	0.71	0.57
5.	0.33	0.74	0.60
6.	0.4	0.72	0.58
7.	0.5	0.65	0.52
8.	0.6	0.53	0.43
9.	0.7	0.39	0.31
10.	0.8	0.25	0.20
11.	0.9	0.18	0.15

For determining conditional formation constant of the thorium (IV)-TOPO system with 3,5 di Br PADAP complex using Likussar and Boltz method the same date in Job's method used. It can be seen from figure that the Job's maximum occurs when the mole fraction of Th(IV) is 0.33. The total amount of thorium corresponding to 0.33 mole fraction is equal to that present in 3.3 ml of 0.0005 M Th(IV). By computing from Beer's law curve, the absorbance of the same amount of thorium is found to be 1.02. Which is ABS-max at 575 nm. The absorbance values are then divided by the ABS-max value at the respective wave length to

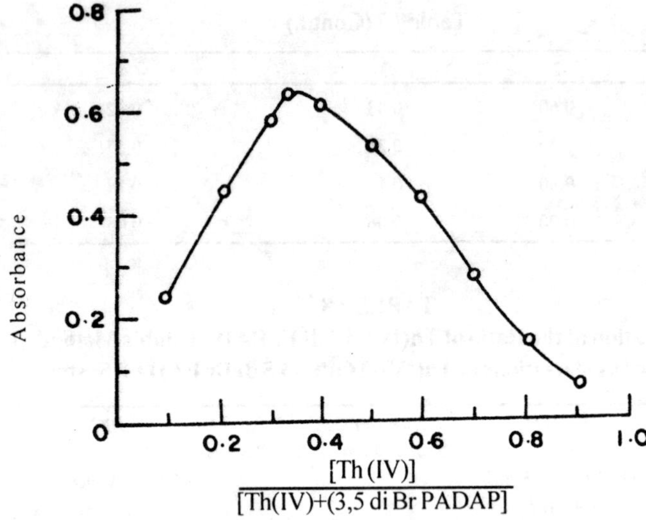

Fig. 7.8 : Job's method of continious variations, determinations of composition of Th(IV) to 3.5 di Br PADAP.

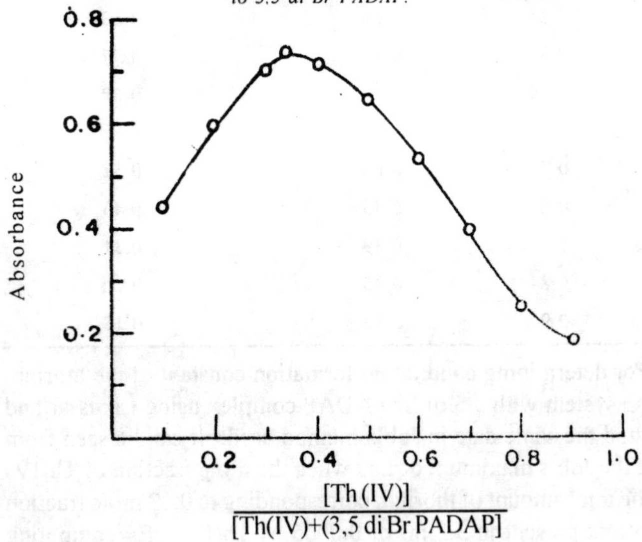

Fig. 7.9 : Job's method of continious variations. determinations of composition of Th(IV) to 3.5 di Br PADAP.

get the Y scale values and are given in Tables. From the table the Y_{max} value is found to be 0.63. The conditional formation of the complex is calculated by substituting the Y_{max} value is found to be 0.63. The conditional formation of the complex is calculated by substituting the Y_{max} value and the value of $K=C_M+C_R$. (Total concentration of the metal + total concentration of the reagent) = 1×10^{-3} in the equation :

$$\log K \frac{\text{Th-TOPO}}{\text{Th-TOPO-3,5 di Br PADAP}} = \frac{0.3552\text{-}2\log K + \log Y_{max}}{3 \log (1-Y_{max})}$$

Similarly from Table Y_{max} and K values are calculated and substituted in the above equation. The average conditional formation constant Log $K \frac{\text{Th-TOPO}}{\text{Th-TOPO-3,5 di Br PADAP}}$ is found to be 7.3 with the data obtained.

References

1. Misciatelli P., Gazz., *Chim. Ital.*, 60, 883, 1949.
2. Bock R. Bock, E. *Naturwis*, 36, 344, 1969.
3. Genov. L., *Monatsh. Chem.* 7, 251, 1959.
4. Bock R. Bock, E. *Z. Anorg. Allg. Chem.* 263, 146, 1950.
5. Banks C.V., Byrd Ch. H. *Anal. Chem.* 25, 416, 1953.
6. Banks C.V. Klingman D.W., Byrd Ch. H. *Anal. Chem.* 25, 992, 1953.
7. Banks C.V. Edward R.E., *Anal. Chem.* 27, 947, 1955.
8. Algerti G. Bettinali C., Salvetti F., Santoli S., Proc. U.N. *Int. Conf. Peaceful uses of At. Energy* 2nd, Geneva, 3, 565, 1958.
9. Levine H. Grimaldi F.S., *Geochim. Cosmochinm. Acta*, 14, 93, 1958.
10. Holcomb H.P. Yoe J.H. *Mocro Chem.* J. 4, 463, 1960.
11. Ishimori T. Data of Inorganic *Solvent extraction Res. Pept. Japan. At. Energy Res. Inst.* JAERI 1047, 1963.
12. Alberti G. Bettinalic C. Salvetti F., Santoli S., *Ann. Chim* (Rome) 49, 199, 1959.
13. Ichinose N., *Talanta* 18, 21, 1971.
14. Ishibashi M., Higashi S., *Bunseki Kagaku* 7, 441, 1958.
15. Ishibashi M., Higashi S., *Bunseki Kagaku*, 4, 14, 1955.
16. Seidi K., Beranak, M. *Chem. Listy* 52, 337 1958.
17. Seidi K. Beranak M., *Collect. Czech. Chem. Commun.* 24, 298, 1959.

18. Fomin V.V. Mairova E.P. *Zh. Neog. Khim.* 1, 1703, 1956.

19. Fomin V.V. Maiorova E.P. *Zh. Neorg. Khim* 1, 2749, 1956.

20. Siekierski S., *J. Inorg. Nucl. Chem.*, 24, 205, 1962.

21. Everest Martin J.V., *Analyst*, 84, 312, 1959.

22. Maiorova E.P. Fomin V.V. *Zh. Neorg. Khim.*, 4, 2511, 1959.

23. Yagi I., Katsumata S., *Kogyo Kagaku Zasshi*, 63, 1926, 1960.

24. T. Ishimori Akatsu *J. Radio Chim. Acta.*, 1, 11, 1962.

25. Hesford E., Muckay H.A.C., Scorgill D., *J. Inorg. Nucl. Chem.*, 4, 315, 1957.

26. Alocock, K. Best G.F., Hesford E., Mckay A.C., *J. Inorg. Nucl. Chem.*, 6, 328. 1958.

27. Hesford E. Mckay H.A.C., *Transfaraday Soc.*, 54, 573, 1958.

28. Invov V.P., Tikhomirov V.I. *Radiokhimiya* 5, 559, 1963.

29. Yoshida M. *Bunseki Kagaku*, 12, 169, 1963.

30. Zolotov Yu. A., Petrukhin O.M., Alimarin I.P. *Zh. Anal Khim.*, 26, 347, 1965.

31. Gerontpulos P., Rigali L., *Radio Chim. Acta.* 3, 122, 1964.

32. Gerontopulos P. Rigali L. Barbanop. G., *Radio Chim. Acta* 4, 224, 1965.

33. Koch. G., *Kerntechnik* 7, 394, 1965.

34. Sato T. Good M.L.U.S., *At. Energy Commi.* ORO - 2576, 23, 1966.

35. Sato T. Good, *M.L.J. Inorg. Nucle. Chem.* 28, 2733, 1966.

36. Sato T., Good M.L. *Solvent Extr. Chem., Proc. Int. Conf., Gotebarg*, 477, 1966.

37. Zhivopistsev V.P., Pyatosin L.P. UchZap., *Perm. Gos. Univ. No.* 159, 254, 1966.

38. Otomo M. *Nippon Kagaku Zasshi* 89, 503, 1968.

39. Zhivopistsev V.P. Pyatosin L.P. *Uch. Zap Perm. Gos. Univ.*, 178, 168, 1968.

40. Kirkbright G.F. Rao A.P., West T.S., *Spectro Sc. Lett.* 2, 69, 1969.

41. Kimura K. *Bull. Chem. Soc.*, Japan, 33, 1038, 1960.

42. Kimura K. *Bull. Chem. Soc.*, Japan, 34, 63, 1961.

43. Dyrssen D. Liem D.H. Acta *Chem. Scand.*, 218, 224, 1964.

44. Byknovtsov V.L. *Radio Khimiya*, 12, 539, 1970.

45. Sheko Z.A. Sinyavakaya E.I., *Radio Khimiya*, 5, 485, 1963.

46. Sheka Z.A., Kriss E.E., Sinyavskaya E.I., *Redkozem Elementary Akad. Nauk. SSSR. Inst. Geokhim. Anal. Khim.* 240, 1963.

47. Stray J. Hladsky E. *Anal. Chim. Acta*, 28, 227, 1963.

48. Newman L. *J. Inorg. Nucl. Chem.* 25, 304, 1963.

49. Newman L. Koltz. P., *J. Phys. Chem.* 67, 205, 1963.

50. Catino. A. Grazz. *Chim. Ital* 89, 939, 1959.

51. Collee R. Rev. *Universelle Mines*, 21, 330, 1965.

52. Day R.A. Jr. Staughtan R.W. *J. Amer. Chem. Soc.* 72,5662, 1950.

53. Hogemann F. *J. Amer. Chem. Soc.* 72, 768, 1950.

54. Meinke W.W. Anderson R.E. *Anal. Chem.* 24, 708, 1952.

55. Waggener W.C., Stoughton R.W. *J. Phys.* 56, 1, 1952.

56. Healy *T.V.J. Inorg. Nucl. Chem.* 19, 314, 1961.

57. Sekine T. Dyrssen D. *J. Inorg. Nucl. Chem.* 29, 1457, 1967.

58. Genov. L. Kasabov G. *Monatsh. Chem.* 100, 594, 1969.

59. Awwal M.A. Shamsuddin A.K. *Nucleus* (Karachi) 7, 190, 1970.

60. Dyrssen D., Svensk. *Kem. Tidskr.*, 65, 43, 1953.

61. Dyrssen D., Dahlberg V., *Acta Chem. Scand.* 7, 1186, 1953.

62. Dyrssen D., Dyrssen M., Johansson E. *Acta. Chem. Scand* 10, 341, 1956.

63. Hogiwara Z., *Technol Rept.*, Tohoku. Univ. 18, 16, 1953.

64. Fritz J.S., Richard M.J., Bystroff A.S., *Anal. Chem.* 29, 577, 1957.

65. Dyrssen D., *Acta. Chem. Scand.* 10, 353, 1956.

66. Saed G.A., Rydenko N.P., Lapttsky A.V., *Z. Anorg, Allg. Chem.* 347, 327, 1966.

67. Alimarin I.P., Chieh Y.H., Vestn. Mosk. Univ. Ser(II) 15, 53, 1960.

68. Rudenko N.P. Saed A.G., Lapitskii A.V., *Radiochimiya*, 7, 142, 1965.

69. Rudenko N.P., Alsamara L., *Vestn. Mosk. Univ. Khim.* 24, 77, 1969.

70. Alimarin I.P. Golovina A.P. Kuteinikov. A.F., Byull. Nauch. *Tekh. Inform., Ministerstvo. Geol. Okhrany Nedr.* SSSR 61, 1957.

71. Fontaine. R., Santoni B. Pommier. C., Guiochon G. *Anal. Chim.* Acta, 62, 337, 1972.

72. Gerkhardt L.I. *Zh. Anal. Khim.* 14, 434, 1959.

73. Kuznetsov V.I. Fang M.O., *Zh. Neorg. Khim.* 5, 1375, 1960.

74. Zharovski F.G., Ostrovskaya M.S. Sukhomlin R.I., Izv. Vyssh. *Ucheb Zaved Khim. Tekhnol.* 10, 989, 1967.

75. Tsutsumi K., *Bunseki Kagaku* 13, 231, 1964.

76. Tsutsumi K., *Bunseki Kagaku* 13, 235, 1964.

77. Sangal S.P. *Bull. Chem. Soc.* Japan, 38, 141, 1965.

78. Lott P.F. Cheng K.L., Kwan B.C. *Anal. Chem.* 32, 1702, 1960.

79. Korkisch J. Janauer G.E., *Mickro. Chim. Acta.* 880, 1961.

80. Korkisch J. Janauer G.E., *Anal. Chem.* 33, 1930, 1961.

81. Budesinsky B. *Collection Czech. Chem. Commun.* 27, 226, 1962.

82. Mukherji A.K., *Micro. Chem.* J. 11, 243, 1966.

83. Budesinsky B.W. Svec. *J. Anal. Chim. Acta.* 61, 465, 1972.

84. Vasilenko V.D., Shanya M.V., *Zh. Analit. Khim.* 20, 636, 1965.

85. Adam J. Pribil R. *Talanta* 16, 1596, 1969.

86. Ishida R., *J. Chem. Soc.* Japan *Pure Chem. Sect.* 86, 1169, 1965.

87. Etimova B., *Anal. Chim. Acta.* 68, 222, 1974.

88. Menis C., Manning D.L., Goldstein D., *Anal. Chem.*, 29, 1426, 1957.

(*References concerned to chapters 6 and 7*)

8

Analytical Applications of Th(IV) TOPO-3,5-diBr-PADAP Ternary System

8.1 Introduction

Though a number of spectrophotometric methods for thorium are reported as reviewed in earlier pages, only few possess practical utility. Even those methods normally practised also have certain limitations. For example, the red complex formed by thorium with xylenol orange[1] has been used for spectrophotometric determination of thorium, but the method is unselective due to interferences from rare-earths and other trivalent metals. By studying the reaction at pH 2.0, the problem was solved but the procedure lacks sensitivity. The sensitivity has been improved by extracting the xylenol orange complex into butanol in the presence of 1,3-diphenyl guanidine[2] but the selectivity is not improved.

In the spectrophotometric determination of thorium with chromoazural S[3] and Cetyl trimethyl ammonium bromide full colour development requires atleast 45 minutes. Serious interferences were caused by Cu(II), Fe(III), and Al(III). One of the most suitable reagents for thorium is thoron suggested by Kuznetsov[4] and investigated by others considerably. Thoron which is benzene-2-arsonic acid 1-azo-1-2-hydroxy naphthalene-3, 6-disulphononic acid permits determination of thorium with great sensitivity in presence of certain associated elements. A 0.2 - 0.3 M HCl medium is optimum. The absorbance decreases with increase in acidity. The molar absorptivity of the complex[5] in 0.25 M HCl is 1.7 x 10^4 l.mol.$^{-1}$ cm^{-1} at λ_{max} 540 nm. Fluoride, oxalate, phosphate and to a lesser degree sulphate interfere. Zirconium and hafnium, titanium, iron (III), U(VI), interfere in this method. Ascorbic acid is used to reduce iron (III) to iron (II) and thus avoids iron interference. The interference of zirconium is reduced with use of mesotartaric acid[5].

Arsenazo III[6] reacts with thorium in strongly acidic medium and is used as a chromogenic reagent for spectrophotometric determination of thorium. With excess of arsenazo III, 2:1 complex with thorium is formed. The maximum absorbance is obtained at 8M HCl, but the reaction in 3 M HCl containing oxalate is more selective. The molar absorptivity of the complex in 3 M HCl is 1.15 x 10^5 $l.mol.^{-1}cm.^{-1}$ at λ_{max} 655 nm. In the presence of oxalic acid thorium can be determined when present along with zirconium, hafnium and niobium. Titanium causes serious interference. Fluoride, phosphate and sulphate interfere in the reaction with Arsenazo III.

Arsenazo I[7] is another azo reagent fairly often used for determining thorium. This reagent is more sensitive but less selective, λ_{max} is observed at 575 nm for this complex.

In the author's investigations it has been observed that Th(IV)-TOPO-3,5-diBr-PADAP method can be advantageously applied for the determination of thorium at trace level when present in various complex matrices. This method has dual advantages where TOPO selectively extracts thorium from high acid medium when present along with various interfering ions, and 3,5-diBr-PADAP imparts necessary sensitivity to the colour reaction making it useful for the determination of thorium even when present in low concentrations. The detailed and systematic studies on the applications of thorium-TOPO-3,5-diBr-PADAP method for the precise analysis of Th in various diversified complex materials (both natural and synthetic) are discussed in this chapter. Various advantages of the present method and its limitations are also described.

As the extraction of Th(IV) with TOPO at 5 M nitric acid medium is found to be highly selective the author has tested the applicability of Th-TOPO-3,5-diBr-PADAP method for the direct analysis of trace quantities of thorium in presence of moderately large quantities of Fe, Ca, Mg, Cu, Ti, Rare earths, and Nb which are likely to exist as matrix components.

The application of this method in presence of large concentrations of complexing anions like, phosphate, oxalate, fluoride, and diethyl dithiocarbamate is also tested for examining their utility as masking agents or pre-extracting agents for the separation of matrix components. This has been carried out by analysing varying amounts of thorium in simulated synthetic samples containing varying concentrations of diverse ions adopting the General Extraction Procedure given in 7.5.

The corresponding results are given in Table 8.1 which indicate usefulness of the present method under varied conditions.

TABLE 8.1

Direct Application of Th(IV)-TOPO-3,5-diBr-PADAP method for analysis of thorium in presence of matrix components

S. No.	*Cation / anion*	*Added*	*Th(IV) taken µg*	*Th(IV) found µg*	*% recovery*
1	2	3	4	5	6
1.	Fe (II)	40 mg	69.0	72.2	104.6
		50 mg	92.0	94.6	102.8
		75 mg	115.0	117.8	102.4
		100 mg	111.0	111.2	100.7
2.	Ca (II)	1 mg	46.0	48.2	104.8
		2 mg	69.0	71.3	103.3
3.	Mg (II)	10 mg	23.0	24.6	106.9
		15 mg	69.0	72.4	104.9
4.	Ce (IV)	100 µg	46.0	48.3	105.0
		200 µg	69.0	70.4	102.9
5.	La (III)	90 µg	92.0	93.6	101.7
6.	Y (III)	200 µg	46.0	47.1	102.4
		200 µg	92.0	93.2	101.3
7.	Al (III)	7 mg	46.0	48.9	106.3
8.	Mn (II)	2 mg	23.0	24.1	104.7
9.	Zn (II)	2 mg	46.0	48.4	105.2
10.	Cu (II)	3 mg	23.0	24.3	105.6
11.	Pb (II)	2 mg	23.0	24.2	105.2
12.	As (III)	2 mg	23.0	24.6	106.8
13.	Cr (III)	10 mg	46.0	47.4	103.1
14.	Co (II)	3 mg	92.0	93.8	101.9
15.	Bi (III)	4 mg	69.0	69.9	101.3
16.	Ti (IV)	75 µg	115.0	118.6	103.1

(Contd.)

Table 8.1 (Contd.)

1	2	3	4	5	6
17.	Diethyl dithio carbonate	10 mg	46.0	48.2	104.7
		100 mg	69.0	71.3	103.3
18.	Oxalate	25 mg	23.0	23.8	103.4
		40 mg	115.0	116.7	101.4
19.	Phosphate	30 mg	46.0	48.3	105.0
		80 mg	115.0	116.5	101.3
20.	Fluoride	15 mg	92.0	95.4	103.6

In situation where the matrix components exist in larger concentrations than the levels investigated (Table 6.1) to achieve precise results adopting the present method, a pre-separation step by either bulk extraction of matrix components, leaving thorium in the aqueous phase, or selective extraction and preconcentration of thorium from the matrix components is suggested. After examining the extraction capacity of various solvents and reagents, diethyl dithiocarbamate is recommended for the first operation while mesityl oxide for latter process.

8.2 Bulk Separation of Matrix Components by Pre-extraction with Diethyl Dithocarbamate

It has been observed in the author's investigations that diethyl dithiocarbamate which forms quantitatively extractable chelates does not react with thorium(IV) and Th-TOPO-3,5-diBr-PADAP method can tolerate large concentrations of DDTC (100 mg). This situation has been exploited for the bulk separation of matrix and constituent components, likely to be associated with thorium in alloys, ores and minerals. It has been observed that metal ions like Fe(III), Ni(II), Cr(VI), Mg(II), Pb(II), Mn(II), Bi(III), W(VI), Cu(II), Co(II), Ca(II), V(V), Ti(IV), U(VI), and Nb(V) can be effectively separated either individually or enblock from thorium by extracting them with diethyl dithiocarbamate at pH 1.0 into 5:2 chloroform: acetone mixture.

8.3 Procedure for the Bulk Separation of Diversified Matrix Components by Pre-extraction with Diethyl Dithiocarbamate and Estimation by Th(IV)-TOPO-3,5-diBr-PADAP Method

Transfer an aliquot of complex materials containing thorium (50-200 μg) and diverse ions into a 250 ml. separatory funnel and add

20.0 ml. of distilled water. To this add 100 mg of sodium diethyl dithiocarbamate solid and make-up the volume to 40 ml with distilled water. Equilibrate this solution with 25 ml portions of chloroform: acetone (5:2) solvent mixture. Repeat the extraction twice or thrice by adding 10 mg of DDTC each time until the aqueous phase is clear and colourless.

Reject the organic phase, add one ml of sodium nitrate solution (40%) to the aqueous phase to reduce Cr(VI) or Mn(VII) to Cr(III) or Mn(II) respectively. Estimate Th(IV) present by Th(IV)-TOPO-3,5,-diBr-PADAP by adopting general procedure given in 7.5.

The above procedure has been tested by analysing synthetic simulated samples of composition similar to various types of complex matrices containing mg(II), Mn(II), Zn(II), Cu(II), Fe(III), in large concentrations and Cr(VI), TI(IV), Nb(V), Pb(II), Al(III), As(III), Ce(IV), La(III), Y(III), in trace concentrations nd the results obtained are given in Table 8.2,3,4 and 5.

These results indicate that Th can be precisely and accurately estimated using Th(IV)-TOPO-3,5-diBr-PADAP method after bulk separation of matrix components by sample and rapid pre-extractions using DDTC.

8.4 Selective Separation and Pre-concentration of Th(IV) from Matrix Components by Pre-extraction with Mesityloxide and Precise Estimation by Th-TOPO-3,5-diBr-PADAP Method

The extraction of thorium with mesityloxide (4-methyle-3-pentene-2 one) is effective. The method is based on the quantitative single extraction of thorium nitrate from solutions saturated with aluminium nitrate as saltingout agent, even in presence of relatively large quantities of phosphate. Equilibrium is attained rapidly, while a uniform yield is obtained both for small (down to 10^{-9} gm/ml) and for large quantities of thorium. Uranium is the only element which is completely extracted in a single extraction together with thorium.

TABLE 8.2

Application of Th(IV)-TOPO-3,5-diBr-PADAP method after bulk separation of diversified complex matrix components by Pre-extraction with DDTC. Analysis results of synthetic samples of simulated composition similar to copper base alloy.

S. No.	Composition	Th(IV) added µg	Found* µg	S.D.	% reco-very	C.V
1.	Cu(II) (75mg)+Zn(II) (10mg)+Ni(II) (4.5mg)+ Al(III) (4 mg)+Fe(III) (5 mg)+Pb(II) (1 mg)	161.0	162.5	3.4	100.7	2.12
2.	Cu(II (76 mg)+Zn(II) (8 mg)+Ni(II) (5 mg)+ Al(III) (6 mg)+Fe(II) (3 mg)+Pb(III) (0.5 mg)	115.0	115.4	2.3	100.4	2.01
3.	Cu(II) (78 mg)+Zn(II) (10 mg)+Ni(II) (6 mg)+ Al(III) (3 mg)+Fe(III) (1 mg)+Pb(II) (0.5 mg)	138.0	137.7	4.0	99.8	2.92
4.	Cu(II) (80 mg)+Zn(II) (7 mg)+Ni(II) (7 mg)+ Al (III) (2 mg)+Fe(III) (2 mg)+Pb(II) (0.7 mg)	69.0	69.3	3.4	100.5	4.96

* Average of six determinations.

TABLE 8.3

Application of Th(IV)-TOPO-3,5-diBr-PADAP method after bulk separation of diversified complex matrix components by Pre-extraction with DDTC. Analysis results of synthetic samples of simulated composition similar to magnesium base alloy.

S. No.	Composition	Th(IV) added µg	Found* µg	S.D.	% reco-very	C.V
1.	Mg(II) (80.5 mg)+Mn(II) (7.2 mg)+Al(III) (4.3mg)+Zn(II) (3.6 mg)+Ni(II) (80 µg)+Fe(III) (100 µg)+Ce(IV) (100 µg)+La(III) (100 µg)	138.0	139.5	2.3	101.1	1.65
2.	Mg(II (78.2 mg)+Mn(II) (6.7 mg)+Al(III) (6 mg)+ZnI(II) (4.6 mg)+Fe(III) (0.2 mg) +Ce(IV) (100 µg)+La(III) (50 µg)	115.0	115.7	2.3	100.6	2.04
3.	Mg(II) (75.4 µg)+Al(III) (8.2 mg)+Mn(II) (8.2 mg)+Zn(II) (3.2 mg)+Fe(III) (0.8 mg)+ Ce(IV) (90 µg)	138.0	139.4	3.1	100.9	2.27
4.	Mg(II) (76.6 mg)+Mn(II) (7.2 mg)+Al(III) (5.5 mg)+Zn(II) (2.0 mg)+Fe(III) (10 mg) +Ce(IV) (25 µg)+La(III) (50 µg)	115.0	116.9	2.5	101.7	2.20

* Average of six determinations.

TABLE 8.4

Application of Th(IV)-TOPO-3,5-diBr-PADAP method after bulk separation of diversified complex matrix components by Pre-extraction with DDTC. Analysis results of synthetic samples of simulated composition similar to Nickel base alloy.

S. No.	*Composition*	*Th(IV) added μg*	*Found* μg*	*S.D.*	*% recovery*	*C.V*
1.	Ni(II) (70 mg)+Cr(III) (3 mg)+Al(III) (6 mg) Mo(VI) (4 mg)+Cu(II) (8 mg)+Ti(IV) (400 μg) Fe(III) (5 μg)+Ce(IV) (80 μg)	70.0	73.7	1.5	105.3	2.15
2.	Ni(II) (80 mg)+Cr(III) (4 mg)+Al(III) (5 mg) Fe(III) (5 mg)+Cu(II) (6 mg)+Nb(V) (600 μg) +Ti(IV) (500 μg)+Y(III) (70 μg)	92.0	92.7	2.0	100.7	2.24
3.	Ni(II) (78 mg)+Cr(III) (5 mg)+Al(III) (2 mg) Fe(III) (8 mg)+Cu(II) (7 mg)+Nb(V) (60 μg)+ Ti(IV) (200 μg)+Ce(IV) (60 μg)	115.0	116.7	1.7	101.4	1.15
4.	Ni(II) (82 mg)+Cr(III) (6 mg)+Al(III) (3 mg) Fe(III) (3 mg)+Cu(II) (2 mg)+Nb(V) (600 μg) +Ti(IV) (200 μg)+La(III) (50 μg)	92.0	92.5	2.3	100.5	2.53

* Average of six determinations.

TABLE 8.5

Application of Th(IV)-TOPO-3,5-diBr-PADAP method after bulk separation of diversified complex matrix components by Pre-extraction with DDTC. Analysis results of synthetic samples of simulated composition similar to Steels.

S. No.	*Composition*	*Th(IV) added μg*	*Found* μg*	*S.D.*	*% recovery*	*C.V*
1.	Fe(III) (90 mg)+Mo(VI) (90μg)+W(VI) (200 μg) +CO(II) (300 μg)+Ti(IV) 80 μg)	46.0	46.1	1.2	100.2	2.50
2.	Fe(III) (92 mg)+As(II) (90 μg)+V(VI)(60 μg) +Mo(VI) (190 μg)+W(VI) (400 μg)	69.0	69.4	1.1	100.6	1.60
3.	Fe(III) (94 mg)+Co(II) (300 μg)+Al(III) (400 μg) +Mo(VI) (0.5 mg)+Co(II) (0.6 mg)	92.0	91.8	2.1	99.8	2.32
4.	Fe(III) (96 mg)+Al(III) (0.5 mg)+Mo(VI) (0.2 mg) +Nb(V) (25 μg)+Co(II) (0.5 mg)	115.0	114.8	1.3	99.8	1.20

* Average of six determinations.

While zirconium is also extracted quite appreciably (about 70 percent and vanadium to the extent of 9 percent), in the back stripping step of thorium with water, zirconium, vanadium and aluminium (in trace quantities) will remain in the organic phase. The extraction process takes about 10 minutes for a single sample. Separation from rare earths is attained by extraction of thorium nitrate with mesityl oxide. Small quantities of rare earths which accompany thorium through mechanical occlusion are separated from the organic solvent by a 3-fold washing with aluminium nitrate solution. The acid concentration in the solution before extraction should be 25 percent by volume. In the determination of thorium in samples containing aluminium as matrix components, thorium is separated from the matrix using mesityloxide in the presence of $LiNO_3$ as salting out agent (Table 8.10)

8.5 Procedure for the Selective Separation and Pre-concentration of Thorium by Extraction with Mesityloxide and Estimation of Th(IV) by Th(IV)-TOPO-3,5-diBr-PADAP Method

Transfer an aliquot of complex material containing thorium (40-200 μg) and diverse ions into a 200 ml beaker, add 2.5 ml of conc. nitric acid so that the total acidity is 2 M and make up the volume to 20.0 ml. Add 19 gms of aluminium nitrate anhydrate and warm to bring that into the solution, then cool and transfer into a 125 ml separatory funnel. Add 20 ml of mesityl oxide to the separatory funnel, shake vigorously for about 1 minute, allow the layers to separate, reserve the organic layer. Repeat the extraction to the aqueous layer with 10 ml of mesityl oxide. Combine the mesityl oxide fractions and discard the aqueous fraction. Strip the rare earths if any, from meistyl oxide by shaking for 1 minute with separate 20 ml portions of aluminium nitrate solution (9.5 gms of aluminium nitrate anhydrate to 5 ml of 2 M nitric acid) rejecting each aqueous phase. Next strip the thorium from mesityl oxide with two 20 ml portions of water. Combine the aqueous phases, reduce the volume to 20 ml by boiling and then estimate thorium by adopting Th(IV)-TOPO-3,5-diBr-PADAP method given in 7.5.

For solutions containing large concentrations of aluminium (as matrix like aluminium alloy), lithium nitrate (16 gms) can be salting-out agent instead of aluminium nitrate.

The above procedure has been tested by analysing synthetic simulated samples of compositions similar to various types of complex matrices containing Mg, Mn, Zn, Ni, Cu, Al, Fe in large concentrations and rare earths [Ce(IV), La(III), Y(III)], Zr(IV), Ti(IV), Nb(V), Pb(II), etc. in trace concentrations and the results are given in Table No. 8.6,7,8,9 and 10.

These results indicate that Th can be precisely and rapidly estimated by Th(IV)-TOPO-3,5-diBr-PADAP method after very effective separation and pre-concentration of thorium from matrix components, by simple clear and effective extraction using mesityl oxide.

TABLE 8.6

Application of Th(IV)-TOPO-3,5-diBr-PADAP method after separation from diversified complex matrix components by Pre-extraction and concentration of Th(IV) with mesityloxide

Analysis results of synthetic samples of simulated composition similar to copper base alloy

S. No.	*Composition*	*Th(IV) added µg*	*Found* µg*	*S.D.*	*% recovery*	*C.V*
1.	Cu(II) (73.7 mg)+Zn(II) (11.4 mg)+Ni(II) (6.7 mg) +Al(III) (2.4 mg)+Fe(III) (4.5 mg)+Pb(II) (1.2 mg)	138.0	138.3	4.4	100.3	3.20
2.	Cu(II) (67.9 mg)+Zn(II) (9.7 mg)+Ni(II) (5.8 mg) +Al(III) (4.6 mg)+Pb(I) (0.7 mg)+As(III) (100 µg)	207.0	208.1	4.9	100.5	2.41
3.	Cu(II) (76 mg)+Zn(II) (7.2 mg)+Ni(II) (4.5 mg) +Al(III) (3.7 mg)+Pb(II) (0.7 mg+Sn(IV) (10 µg)	115.0	115.4	2.5	100.4	2.20
4.	Cu(II) (81.7 mg)+Zn(II) (8.2 mg)+Ni(II) (5.2 mg) +Al(III) (1.2 mg)+Pb(II) (0.9 mg)+As(II) (20 µg)	46.0	47.2	0.6	102.8	1.40

* Average of six determinations.

TABLE 8.7

Application of Th(IV)-TOPO-3,5-diBr-PADAP method after separation from diversified complex matrix components by Pre-extraction and concentration of Th(IV) with mesityloxide

Analysis results of synthetic samples of simulated composition similar to magnesium base alloys

S. No.	*Composition*	*Th(IV) added µg*	*Found* µg*	*S.D.*	*% recovery*	*C.V*
1	2	3	4	5	6	7
1.	Mg(II) (76.7 mg)+Mn(II) (6.4 mg)+Al(III) (3.7 mg) +Zn(II) (6.2 mg)+Ni(II) (80 µg)+Fe(III) (120 µg) +Ce(IV) (120 µg)+La(III) (120 µg)	115.0	115.2	2.4	100.2	1.41

(Contd.)

Table 8.7 (Contd.)

1	2	3	4	5	6	7
2.	Mg(II) (75.4 mg)+Al(III) (6.7 mg)+Zn(II) (6.2 mg) +Mn(II) (7.3 mg)+Ce(IV) (80 μg)+Cu(II) (12 μg)	69.0	71.4	0.74	103.4	1.02
3.	Mg(II) (82.4) mg)+Mn(II) (7.8 mg)+Zn(II) (2.4 mg) +Al(III) (4.4 mg)+Cu(II) (18 μg)+Ce(IV) (200 μg)	146.0	148.0	4.7	101.3	3.21
4.	Mg(II) (70.8 mg)+Al(III) (4.7 mg)+Mn (II) (3.4 mg) +Zn(II) (2.6 mg)+Zr(IV) (80 μg)+Ce(IV) (200 μg)	169.0	170.3	2.7	100.8	1.61

* Average of six determinations.

TABLE 8.8

Application of Th(IV)-TOPO-3,5-diBr-PADAP method after separation from diversified complex matrix components by Pre-extraction and concentration of Th(IV) with mesityloxide

Analysis resulis of synthetic samples of simulated composition similar to nickel base alloys

S. No.	*Composition*	*Th(IV) added μg*	*Found* μg*	*S.D.*	*% recovery*	*C.V*
1.	Ni(II) (74 mg)+Al(III) (7.2 mg)+Cu(II) (8.2 mg)+ Mo(VI) (40 mg)+Ti(IV) (400 μg)+Ce(IV) (180 μg)	123.0	122.8	2.6	99.8	2.20
2.	Ni(II) (81.3 mg)+Cr(III) (4 mg)+Fe(III) (3.4 mg) Cu(II) (7.2 mg)+Nb(V) (800 μg)+Ti(IV) (300 μg) +La(III) (144 μg)	146.0	146.5	4.2	100.3	2.92
3.	Ni(II) (78 mg)+Al(III) (2.4 mg)+Fe(III) (3.4 mg) Cu(II) (7.4 mg)+Ce(IV) (120 μg)+Nb(V) (60 μg)	169.0	169.8	4.8	100.4	2.80
4.	NI(II) (81.4 mg)+Cr(III) (2.7 mg)+Al(III) (3.7 mg) +Fe(III) (7.6 mg)+Cu(II) (2.7 mg)+Ti(IV) (250 μg)	176.0	179.9	3.8	102.2	2.10

* Average of six determinations.

TABLE 8.9

Application of Th(IV)-TOPO-3,5-diBr-PADAP method after separation from diversified complex matrix components by Pre-extraction and concentration of Th(IV) with mesityloxide

Analysis results of synthetic samples of simulated composition similar to steels

S. No.	*Composition*	*Th(IV) added μg*	*Found* μg*	*S.D.*	*% recovery*	*C.V*
1	2	3	4	5	6	7

(Contd.)

Table 8.9 (Contd.)

1	2	3	4	5	6	7
1.	Fe(III) (95 mg)+Mo(VI) (0.2 mg)+Co(II) (0.2 mg) Ti(IV) (60 μg)+Ce(IV) (40 μg)+La(III) (10 μg)	169.0	171.2	3.8	101.3	2.21
2.	Fe(III) (96 mg)+V(VI) (64 μg)+As(III) (100 μg) +Mo(VI) (250 μg)+Ce(IV) (60 μg+La(III) (30 μg)	92.0	92.2	1.8	100.2	2.01
3.	Fe(III) (97 mg)+V(V) (100 μg)+M(IV) (400 μg) +Mo(VI) (200 μg)+Co(II) (0.6 mg)	115.0	115.4	2.6	100.4	2.30
4.	Fe(III) (98 mg)+Al(III) (0.5 mg)+Nb(V) (25 μg) +Mo(VI) (1.2 mg)+Ce(IV) (100 μg)+La(III) (50 μg)	146.0	145.1	2.8	99.4	1.90

* Average of six determinations.

TABLE 8.10

Application of Th(IV)-TOPO-3,5-diBr-PADAP method after separation from diversified complex matrix components by Pre-extraction and concentration of Th(IV) with mesityloxide

Analysis results of synthetic samples of simulated composition similar to aluminium alloys

S. No.	*Composition*	*Th(IV) added μg*	*Found* μg*	*S.D.*	*% recovery*	*C.V*
1.	Al(III) (74 mg)+Cu(II) (3.8 mg)+Mg(II) (4.8 mg) +Fe(III) (7.2 mg)+Mn(II) (2.4 mg)+Ce(IV) (600 μg) +Zr(IV) (40 μg)	46.0	47.9	0.7	104.2	1.40
2.	Al(III) (72.3 mg)+Cu(II) (7.8 mg)+Mg(II) (5.2 mg) +Fe(III) (4.8 mg)+Mn(II) (3.8 mg)+U(VI) (50 μg) +Ti(IV) (100 μg)+Ce(IV) (250 μg)	115.0	118.2	0.7	102.8	0.61
3.	Al(III) (81.9 mg)+Mg(II) (3.4 mg)+Cu(II) (8.7 mg) +Cr(III) (0.6 mg)+Mn(II) (0.6 mg)+Ti(IV) (250 μg) +La(III) (64 μg)	138.0	141.5	0.9	102.5	0.67
4.	Al(III) (7.6 mg)+Mg(II) (6.2 mg)+Cu(II) (2.4 mg) +Mn(II) (1.2 mg)+Ti(IV) (300 μg)+Ce(IV) (120μg) +Nb(V) (26 μg)	69	72.3	0.7	104.8	1.0

* Average of six determinations.

However, in actual analysis of thorium using Th(IV)-TOPO-3,5-diBr-PADAP method in various natural and industrial materials composed of diversified matrix components a proper combinations of these two processes i.e.,

(a) Bulk separation of matrix component by pre-extraction with DDTC, or aliquot 336 for Zirconium and titanium.

(b) Selective separation of thorium by extraction with mesityl oxide from matrix components.

Makes the method more precise, adoptable and selective as these two separation processes are simple, rapid and very efficient.

8.6 Application of Th-TOPO-3,5,diBr-PADAP Method for the Determination of Th in Titanium and Zirconium Matrices

Th(IV)-TOPO-3,5-diBr-PADAP is a sensitive and selective method for spectrophotometric determination of thorium, but high concentrations of titanium(IV) and zirconium(IV) cause serious interference which prevents direct applications of Th(IV)-TOPO-3,5-diBr-PADAP method in titanium and zirconium concentrates. Further the matrix separation procedures with DDTC or thorium pre-concentration method with mesityl oxide do not work effectively particularly with large concentration of Zr and Ti.

A simple separation of thorium before its determination was reported[8] in which iron, uranium, zirconium and other metal ions were removed by extraction with alamine-336 (a long chain tertiary amine) from 9 M hydrochloric acid. Unfortunately most of the titanium remains in the aqueous phase with thorium and despite modifications the amount of titanium that can be tolerated in the sample is severely limited.

Aliquot-336, a long chain quaternary, was found to be much more efficient than alamine-336 for the extraction of titanium(IV) from hydrochloric acid media (87% extraction from 10 M Hydrochloric acid) and its use allows thorium to be determined in rutile and titanium concentrates. In addition, the extraction coefficient of 150 for zirconium (from 10 M hydrochloric acid) is sufficiently high for most of zirconium to be extracted into the organic phase. The interference of remaining traces of zirconium and titanium can be removed by adopting selective extraction of thorium with mesityl oxide procedure.

8.7 Procedure for the Bulk Separation of Matrix Components (Fe, Zr and Ti) by Pre-extraction with Aliquot-336

Transfer an aliquot of complex material containing thorium (50-200 mg) and matrix components into a 125.0 ml separatory funnel and add conc. HCl so that total acidity is 10 M. Make up the aqueous layer to

25.0 ml. of aliquot-336 in cyclohexane and shake for two minutes. Repeat this extraction with another 25.0 ml of aliquoat 336 nd reject the organic layer. Boil the aqueous layer to near dryness. Add distilled water and make up the volume to 20.0 ml. Extract thorium with 20.0 ml portions of mesityl oxide (to separate thorium from any traces of zirconium or titanium) and then back extract it by shaking with distilled water and estimate thorium using Th(IV)-TOPO-3,5-diBr-PADAP method given in 7.5.

The above procedure has been tested by analysing synthetic simulated samples containing zirconium, titanium, uranium and iron in different concentrations.

These results indicate that Th can be precisely and rapidly estimated by Th(IV)-TOPO-3,5-diBr-PADAP method after very effective separation of matrix components from thorium (by simple clear and effective extraction using aliquot-336) and later pre-concentration of thorium with mesityl oxide.

8.8 Application of Th(IV)-TOPO-3,5-diBr-PADAP method for the analysis of natural materials

(a) Analysis of monazite sand after adopting pre-extraction and separation method with mesityl oxide

Procedure

Decompose the weighed sample (0.5-1.0 gm) of the monazite sand by evaporation with sulfuric acid on a sand bath, add 200 ml of water and 5 ml of concentrated sulfuric acid to the residue and filter. In the filtrate precipitate the hydroxide with aqueous ammonia. Wash the collected precipitate with aqueous ammonia nitrate solution and dissolve in 1:4 hydrochloric acid. Dilute with water to 200 ml, add 50 ml of concentrated hydrochloric acid, warm and add 20 ml of 10% oxalic acid solution. Wash the precipitated oxalates with 1:1 mixture of 1:10 hydrochloric acid and 1% oxalic acid solution. Filter the thorium oxalate through a glass fiiter and wash with 2% oxalic acid and reject the filtrate. Place 100 ml beaker under the funnel and dissolve the thorium oxalate precipitate by 3 portions of 10 ml each hot 1:1 HNO_3 following each portion with a limited quantity of hot water. Reject the residue on filter paper. Evaporate the solution to dryness. Add 3-4 ml fuming HNO_3, cover the beaker and allow to stand for a few minutes on the steam bath. Remove the cover and evaporate the solution to dryness. Dissolve the residue in 10 ml of 15:85 HNO_3 and heat the solution to dissolve the residue. A certain quantity of this may sometimes remain undissolved, but dissolution occurs after addition of aluminium nitrate.

Add 19 gm $(Al(NO3)_3 . 9H_2O$ and heat the mixture until the crystals dissolved. Allow to cool. Place the solution in a 125.0 ml separatory funnel. Add 20.0 ml mesityl oxide to the separatory funnel. Shake for 1 minute. After the layers are separated, equilibrate the aqueous layer again with another 20.0 ml of mesityl oxide. Combine the mesityl oxide layers, rejecting the aqueous layer.

Treat the organic layer with 3 separate portions of 20 ml of aluminium nitrate (Washing solution containing 9.5 gms of $Al(NO_3)_3$ in 5 ml of 15:85 HNO_3. Reject the aqueous portions. Remove the thorium from the organic solvent with 2 portions of water of 20 ml each and transfer the aqueous layers into a 150 ml beaker.

Add 1 ml of conc. HNO_3. Reduce the Volume of the aqueous layer to 80 ml. Transfer it into a 100 ml volumetric flask and make up to 100 ml with distilled water. Estimate thorium adopting General Extraction Procedure given in 7.5.

The results of author's investigations on the analysis of monazite sand samples by adopting the above procedures are given in Table No. 8.11 which indicate applicability of the present method for the analysis of monazite sand.

8.8 (b) Analysis of Rutile and Ilmenite Samples after Bulk Separation of Matrix Components by Pre-extraction with Aliquat-336

Procedures

Weigh 0.5 gm of concentrate sample (200 mesh) into a platinum crucible. Add 2.5 gm of sodium peroxide and mix. Sinter at 480 ± 20°C for one hour. Dissolve the sinter in 150 ml of water and boil the solution for 15 minutes and cool. Filter (Whatman 41) and wash the precipitate with 500 ml of 2% sodium hydroxide solution, then dissolve it in 60 ml of hot 8 M hydrochloric acid. Boil the solution until the volume is reduced to 45 ml and then cool. Transfer it into a 50 ml volumetric flask and dilute to the mark with 6 M hydrochloric acid.

Transfer 5.0 ml of sample solution (6M in hydrochloric acid) to 60 ml separatory funnel and add 15.0 ml of hydrochloric acid and mix. add 20 ml of aliquot-336 in cyclohexane and shake for two minutes. Reject the organic layer. Repeat the extraction second time with another 20.0 ml of 10% aliquat-336 in cyclohexane. Retain the aqueous phase. Evaporate the solution to dryness and add 10 ml of distilled water. Transfer the aqueous phase into 125 ml separatory funnel and add sufficient HNO_3 to keep the overall acidity at 2M. Then Pre-concentrate Th by extracting with mesityl oxide following the procedure given in 8.5, and then estimate thorium by Th(IV)-TOPO-3,5-diBr-PADAP method given in 7.5

The applicability of the above procedure has been tested by analysis of actual rutile and ilmenite samples (Table 8.11) which yielded reasonably accurate results with standard deviations in the range 1.9 to 2.8.

TABLE 8.11

Application of Th(IV)-TOPO-3,5-diBr-PADAP Method After Bulk Separation of Diversified Matrix Components

Analysis Results of Natural Samples

S. No.	*Samples*	*Th given (%)*	*Th added (%)*	*Found* %*	*% recovery*	*S.D*
1.	Monazite Sand	7.80	–	7.82	100.5	3.9
2.	Ruttile Sample 1	0.044	0.115	0.165	102.5	2.8
3.	Ruttile Sample 2	0.042	0.138	0.189	103.8	2.7
4.	Ilmenite Sample 1	0.01	0.115	0.164	105.1	1.9
5.	Ilmenite Sample II	0.043	0.138	0.184	101.7	2.8
6.	Titanium alloy [NBS SRN 50 (a)]	–	0.200	0.206	103.0	2.9

* Average of six determinations

8.9 Analysis of Th in Synthetic industrial materials

(a) Estimation of Th in Mg alloy After Adopting Bulk Separation of the Matrix Component with DDTC Extraction

Procedure

Place 0.5 gm of the alloy in 400 ml beaker and add 50 ml of distilled water conc. HCl in small portions until all the alloy is completely dissolved. Heat till boiling, cool and if necessary, filter it. Fuse any residue with potassium bisulphate and remove traces of thorium in the melt (extract) with H_2O+HCl. Filter and combine the filtrate in a 250 ml beaker, evaporate to near dryness and add 100 ml of distilled water. Separate the matrix components in bulk by extracting with DDTC and pH 1.0 into 5:2 chloroform: acetone mixture as per the (procedure in 8.3) and then estimate Th by Th-TOPO-3,5,diBr-PADAP method described in 7.5.

Since the Th containing Mg alloys could not be procured by the author, the author tested applicability of the above procedure by analysing actual NBS SRM 171 magnesium base alloy (0.5 gms) after spiking definite amounts of thorium and the corresponding results are given in Table 8.12.

8.9 *(b) Estimation of Th in Steels After Adopting Bulk Separation of the Matrix Component with DDTC Extraction*

Procedure

Weigh 0.5 gm of steel into a platinum dish, add 15 ml of HCl + 5 ml of HNO_3 and heat gently until the sample is dissolved. Add 5 ml of perchloric acid and evaporate the mixture until dense fumes are evolved. Cool the mixture, add drops of 40% HF, 10 ml of nitric acid and heat to boiling. After cooling, make up the volume to 40-50 ml. Filter through a Whatman 40 filter paper into a 100 ml. Volumetric flask. Wash the residue and make up the filtrate to 100.0 ml with distilled water. Transfer an aliquot of the solution into a separating funnel and estimate thorium adopting General Procedure given in 7.5.

Since the author could not procure thorium-containing steels, the author tested the applicability of the above procedure by analysing actual NBS SRM 19g steel sample after spiking known amount of thorium and the corresponding results are given in Table 8.12.

8.9 *(c) Estimation of Th in Aluminium Alloy After Adopting Separation and Pre-concentration of Thorium from Matrix Component with Mesityl Oxide Extraction*

Procedure

Weigh 0.5-1.0 gm of sample of alloy in a platinum or nickel dish provided with lid, add 3 gms of solid granulated NaOH and gradually add water at the rate of 75 ml of each gram of alloy. Heat and boil for 30 minutes. Transfer the contents of dish into a porcelain dish of large size, to which 30 ml of 1:1 sulphuric acid has been previously added for each gm of alloy. Heat the content of dish rapidly with frequent stirring and allow to remain on a sand bath for 30 minutes until strong evolution of sulphuric acid vapours occurs. Cool and add 100 ml of water. Filter the precipitate through whatman filter paper and reject the residue.

Fuse the residue with $KHSO_4$, cool the melt and extract with 100 ml water and two ml of HCl. Reduce the volume of the filtrate to 25 ml by boiling and transfer into 50 ml volumetric flask and make up to the mark with 2M HCl to neutralize. Then estimate Th by the Procedure given in 8.5.

Since Th containing Al alloys are not available with the author, the author tested the applicability of the above procedure by analysing actual NBS SRM 859 aluminium alloy after spiking know amounts of thorium and the corresponding results are given in Table 8.12.

TABLE 8.12

Application of Th(IV)-TOPO-3,5-diBr-PADAP Method After Bulk Separation of Diversified Matrix Analysis Results of Some Spiked Standards (Synthetic Industrial Materials)

S.No.	*Composition*	*Th added (%)*	*Found* %*	*S.D*	*% recovery*
1.	**Magnesium base alloy :** NBS SRM 171:Mn (0.45%), Cu (0.0112%) Ni (0.0009%), Al (2.98%), Pb (0.0033%), Zn (1.05%), Fe (0.0018%),	0.184	0.192	2.3	104.4
2.	**Nickel base alloy :** NBS SRM 867: Mn (0.4%), Cu (1.6%) Ni (43%), Cr (23%), Mo (2.7%), Al (0.06%)	0.175	0.178	2.4	101.7
3.	**Aluminium base alloy :** NBS SRM 859: Mn (0.078%), Cu (1.5%), Ni (0.063%), V (0.0082%), Fe (0.078%), Mg (2.45%), Zn (5.46%)	0.184	0.189	2.5	102.7
4.	**Steel** NBS SRM 19g, Cu (0.093%), Ni (0.066), Cr (0.374%), V (0.012%), Mo (0.013%), Co (0.012%)	0.175	0.181	1.9	103.4
5.	**Titanium base alloy :** NBS SRM 50 (a), Mn (1.60%), Fe (0.19%), Al (2.74%)	0.184	0.191	2.2	103.8

* Average of six determinations.

8.9 *(d) Estimation of Th in Nickel Base Alloy After Adopting Bulk Separation of the Matrix Component with DDTC Extraction*

Procedure

Dissolve 100 mg of sample by heating with a mixture of 10 ml of concentrated nitric acid and 10 ml of concentrated nitric acid and 10 ml of concentrated hydrochloric acid and 10 ml of water. Transfer the solution to a 50 ml volumetric flask and dilute to volume with water. Transfer a suitable aliquot of the solution to a beaker and evaporate the solution to about one ml. Dilute the solution with distilled water to 10.0 ml. Separate the matrix components in bulk by extracting with DDTC at pH 1.0 into 5:2 chloroform: acetone mixture as per the procedure given in 8.5, and estimate Th by adopting Th(IV)-TOPO-3,5-diBr-PADAP method described in 7.5.

The applicability of the above procedure has been tested by analysing actual NBS SRM 867 nickel base alloys after spiking known amounts of thorium and the corresponding results are given in Table 8.12.

8.9 *(e) Estimation of Th in Titanium Base Alloy After Adopting Bulk Separation of the Matrix Component with Aliquot-336 and Pre-concentration of Thorium with Mesityl Oxide*

Procedure

Transfer 0.5 gm of titanium base alloy into a 250 ml beaker, dissolve it in a mixture of 120 ml of 4:1:1 H_2O:HCl:H_2SO_4. When the reaction has ceased and all titanium is dissolved, add enough nitric acid to oxidize the titanium (III) to titanium (IV). The solution changes from purple to colourless. Heat it to fumes of sulphuric acid to remove chlorides and to reduce the volume to 80.0 ml. Place in 100 ml volumetric flask. Separate the matrix components in bulk by extracting with aliquot-336 procedure given in 8.7 and pre-concentrate thorium by mesityl oxide procedure given in 8.5. Then estimate Th adopting Th(IV)-TOPO-3,5-diBr-PADAP general extraction method given in 7.5.

Since there is no availability of thorium-containing titanium base alloy, the applicability to the above procedure has been tested by analysing actual NBS SRM 50 (a) titanium base alloy and the corresponding results are given in Table 8.12.

The critical examination of the analysis results of various synthetic industrial materials of Th indicate that, by proper combination of optimised matrix separation methods, the Th(IV)-TOPO-3,5-diBr-PADAP method can be advantageously applied for the precise determination of thorium at trace level in these complex matrices.

However, in actual analysis of thorium using Th(IV)-TOPO-3,5-diBr-PADAP method in various natural and industrial materials composed of diversified matrix components, a proper combination of the three optimised procedures developed by the author Viz.:

(a) Bulk separation of matrix component by pre-extraction with DDTC.

(b) Selective separation of thorium by extraction with mesityl oxide from matrix components.

(c) Bulk extraction of titanium, iron, uranium and zirconium by extraction with aliquot-336.

make the method more precise, adoptable and give accurate results.

These three selective separation processes are simple, rapid and efficient when compared to other practised methods.

References

1. Otomo. M. *Bunseki Kagaku* 14,229, 1965.
2. Otomo. M. Wakamatsu Y., *Nippon Kagaku Zasshi.* 89,1087, 1968.
3. Etimova B., *Anal. Chim. Acta.* 68,222, 1974.
4. Kuznetsov V.I. *Doklady Akad SSSR.* 50,233, 1945.
5. Palmer, A.R., *Anal. Chim. Acta.* 14,162, 1956.
6. Savvin S.B. *Talanta* 8,673, 1961.
7. Kumetsov V.I. and Nikolskaya I.V., *Zh.. Analit. Khim.* 15,229, 1960.
8. Paklans P. *Anal. Chim. Acta.* 58,463, 1972.

9

Results and Discussion of the Trace Analysis of Thorium

9.1 General Conditions

The major industrial use of Th is as an atomic fuel, for which purpose, it is essential that it should be obtained in a state of high purity, especially with respect to boron, cadmium and the rare earths, all of which are possible impurities. The preparation of very pure metal is a complex problem, in view of high melting point of thorium and the ease with which the heated metal reacts with hydrogen, nitrogen, oxygen and carbon as well as the difficulty of isolating it from its ores free from the many elements with which it is associated in nature, particularly the rare earth.

The high melting point (1650°C-1800°C) and high atomic weight of the metal suggest its use in heat resistant alloys[1]. It forms alloys readily with aluminium[2-7], iron[8], copper[9], nickel and cobalt[10] and magnesium[11-13]. In these materials thorium concentrations determined should always be checked against a second independent analytical method. So there is a need to develop alternative methods for the analytical control of Th which are not only precise but rapid enough to be used in quality control. Preliminary separation of thorium from Zr, Ti and U from each other and from interfering substances is often required before their determination.

Though a number of spectrophotometric methods for thorium are reported, many of these suffer from pronounced drawbacks and lack practical utility. Of the various colorimetric determinations, the sensitive are the reaction between thorium and thoron (I) in acid medium and arsenazo III which reacts with thorium in strongly acidic medium. But in the reaction with thoron I, fluoride, oxalate phosphate and sulphate interfere and Fe(III) should be reduced to Fe(II) to minimize its interference. In the reaction with arsenazo III acidity should be maintained constant both in standards, samples and blank as otherwise non-reproducible results will occur.

Pyridylazo compounds with a hydroxy group in the ortho position with respect to azo group are valuable chromogenic reagents. They have been used widely in complexometric and extraction photometric methods for the determination of many metal ions. Highly sensitive reagents have been obtained by the introduction of diethylamino group which is a strong electron - donating group in the para position next to azo group. The azo dye 2-(2-Pyridylazo)-5 diethylamino phenol (PADAP) and its bromine derivatives offer potential applications very sensitive and selective spectrophotometric reagents for many metal ions.

The colour reaction between thorium and PADAP and 5 - Br PADAP was studied by Gusev and schurova[14]. These authors studied the reaction between PADAP, 5 - Br PADAP and thorium at pH 4.5 - 5 and 3-3.5 respectively in aqueous medium with max at 520 nm. However, these suffer from serious interferences from a number of other cations as they also react similarly with PADAP and 5 - Br. PADAP.

In the author's investigations it has been observed that thorium (IV) can be selectively extracted with TOPO in toluene at 2-7 M nitric acid medium as a colourless complex which is free from interferences from many associated ions like Fe(III), Bi(III), Cu(II), Mo(VI), Mn(II), Ca(II), Mg(II), Pb(II), As(III), Co(II), etc., It is further observed that the selectively extracted colourless Th(IV)-TOPO complex system can be converted into an intensely coloured ternary complex by the reaction with chromogenic complexing agent like 3,5 di Br PADAP in non-aqueous medium. This method has dual advantages where TOPO extracts thorium selectively and 3,5 di Br PADAP imparts necessary sensitivity to the colour reaction making it useful for the determination of thorium even when present in low concentrations.

Th(IV) is quantitatively extracted (99%) from 0.1 M to 3 M nitric acid medium into toluene containing TOPO (0.05 M) while its extraction is slightly lower at higher acidities Ω 95% from 3-7 M HNO_3. But at this high acidity range the extraction of thorium is highly selective in nature and is free from interferences from various associated ions. In HCl medium, also similar extraction of thorium can be achieved. But from $HClO_4$ or H_2SO_4 the extraction is not quantitative and non-reproducible results are obtained.

It is observed from these investigations that among hexane chloroform, carbon tetrachloride, cyclohexane, nitrobenzene, benzene, toluene, methyl isobutyl ketone tested as solvents for the extraction of thorium, toluene, carbon tetrachloride and cyclohexane are found to extract thorium quantitatively. However in the spectrophotometric studies the colour reaction of Th(IV)-TOPO complex with 3,5 di Br PADAP, it is observed that toluene medium as a facile medium for the reaction giving intensely coloured species. A ten minute shaking period is necessary for the quantitative extraction of thorium with 0.05 M TOPO in toluene at 5 M nitric acid medium.

It is observed that chloride (100 mg), nitrate (100 mg), fluoride (15 mg), sulphate (10 mg), phosphate (100 mg), oxalate (50 mg), nitrite (10 mg) ascorbic acid (10 mg), Al(III) (7 mg), Ca(II) (4 mg), Mg(II) (18 µg), La(III) (200 µg), Ce(IV) (200 µg), Y(III) (100 µg) do not interfere in the extraction of thorium with TOPO both in nitric or hydrochloric acid media. Hydrogen peroxide, (one ml of 38 volumes), EDTA (0.5 mg), Ti(IV) (80 µg), U(VI) (45 µg), Zr(IV) (45 µg) and Nb(V) (20 µg), interfere both in nitric acid and hydrochloric acid medium.

However in nitric acid medium effective extraction of thorium (IV) can be achieved even in presence of large concentrations of Cu(II) (3 mg), Zn(II) (3 mg), Mo(V) (1 mg), Fe(III) (100 mg), Bi(III) (4 mg), As(III) (4 mg), Mn(II) (3 mg), V(V) (2 mg) and diethyldithiocarbamate. But when the extraction is carried out in HCl medium it is observed that these ions interfere in the extraction of thorium.

In general it is observed that the extraction of Th(IV) from TOPO in 5 molar nitric acid medium is free from interference from associated substances and tolerance limits of many metals are high when compared to HCl medium. It is observed that a 2000 fold TOPO to thorium (IV) is necessary for quantitative extraction of thorium.

In the experimental conditions studied it is observed that TOPO extracts nitric acid as per the following equilibrium :

$$H^+_{(aq)} + NO_3^- + n\ TOPO_{(0)} \rightleftharpoons (HNO_3nTOPO)_0$$

At low acidities it is found to be a 1:1 adduct. The extraction of thorium (IV) with TOPO at low acidities ≤ 0.4M is found to be fairly independent of acid extraction and equilibrium extraction data suggests the extraction occurs as Th(IV) $(NO_3)_4$ 3 TOPO species represented by the following equilibrium reaction :

$$Th(IV) + 4\ NO_3^- + 3\ TOPO \rightleftharpoons Th(NO_3)_4\ 3\ TOPO$$

However at higher acidities the extraction of thorium is no longer independent of acid extraction and formation of higher adducts like :

$$[Th(IV)NO_3)_4\ n\ HNO_3.\ 3\ TOPO]\ n\ HNO_3\ TOPO$$

are indicated from extraction data.

It is observed in the author's investigations that the colourless Th(IV)-TOPO extracted species can be converted into an intensely coloured one by adding a chromogenic complexing reagent in non-aqueous medium. Of the various chromogenic complexing reagents like 4(2 Pyridylazo) resorcinol (PAR). 1(2 pyridylazo) 2 naphthol (PAN) 2(2 - Pyridylazo) 5-diethyl amino phenol (PADAP) 5 Br PADAP and 3,5 di Br PADAP tested, 3,5 di Br PADAP gave high sensitivity colour reaction with good precision.

Thus extraction of thorium from high acidic medium with TOPO into toluene and addition of 3,5 di Br PADAP in ethanol + pyridine non-awueous medium makes the reaction more selective and sensitive.

It is observed in the present investigations, that the presence of pyridine in the reaction medium accelarates the colour formation of Th(IV)-TOPO with 3,5 di Br PADAP and increases its sensitivity. Of the various bases like pyridine, picoline, triethanolamine, aniline tested for their effect on the colour formation of Th(IV)-TOPO-3,5 di Br PADAP system, pyridine is found to be very effective and pyridine composition from 30 to 50% in the non-aqueous medium yields maximum absorbance with minimum time. Among the various diluents tested such as acetone, absolute ethanol, amyl alcohol, 1,4 dioxane and toluene, absolute ethanol gave maximum absorbance. To get the best results a period of 15 minutes standing time is required after the addition of pyridine and before the addition of absolute ethanol. The working capacity of specific non-aqueous medium (Pyridine + ethanol + toluene) on the colour formation may be due to the fact that solvent properties like dielectric constant, polarity, Polarisability can directly affect the thermodynamics of complex formation reactions through solvent-solvent and solvent-solute interactions.

It is neccssary to add 3,5 di Br PADAP first to the organic extract and after adding pyridine this mixture should be kept standing for 15 minutes. Then the whole mixture should be made upto the volume by adding the diluent. This order should be strictly adhered to attain maximum reproducible colour development. There is no effect of traces of water in ethanol practically on the absorbance of Th(IV)-TOPO-3,5 di Br PADAP complex.

From detailed studies it is indicated that a 30 fold excess of 3,5 di Br PADAP in ethanol is sufficient for attaining maximum absorbance and thereafter addition upto 100 fold has no effect on the absorbance of Th(IV)-TOPO 3,5 di Br PADAP complex.

Th(IV)-TOPO-3,5 di Br PADAP system obeys Beer's law upto 178 μg (Th in the initial aqueous phase which comes to 3.5 μg/ml in the final organic phase) and exhibits molar absorptivity of 1.2 x 10^5 1. mol.$^{-1}$ cm.$^{-1}$ at 575 nm which is higher than those earlier reported with oxine[15] (1.0 X $10^4$1.mol.$^{-1}$cm.$^{-1}$), molybdophosphoric acid[16] (3.3 x 10^4 1.mol.$^{-1}$cm.$^{-1}$) bromopyrogallolred[17] (5.1 x 10^4 1.mol.$^{-1}$cm.$^{-1}$), murexide[18] (1.8 X $10^4$1.mol.$^{-1}$cm.$^{-1}$), methylthymol blue[19], (2.9 X $10^4$1.mol.$^{-1}$cm.$^{-1}$), 4 (2 thiazolylazo) resorcinol[20], (3.15 X 10^4 1.mol.$^{-1}$cm.$^{-1}$) and xylenol orange[21], (5.5 X 10^4 1.mol.$^{-1}$ cm.$^{-1}$).

The Sandell's sensitivity of the colour reaction is found to be 0.00193 μg cm^{-2}. From the Ayre's and Ringbom's plots the optimum range with minimum photometric error for the estimation of thorium is found to be in the range 0.3 μg to 2 μg per ml of thorium.

The precision of the method is evaluated by the analysis of different concentrations of thorium (in the range 27-200 μg in the initial aqueous phase, each analysing six times) which gave results with a relative standard deviation in the range 0.3-3.5 and coefficient of variation ranging from 0.2–2.4

In the detailed studies to examine the interference of foreign ions on Th(IV)-TOPO 3,5 di Br PADAP spectrophotometric method it is observed that the ions which affect the extraction of thorium also interfere in the spectrophotometric method.

Detailed investigations on the spectral characteristics of the Th(IV)-TOPO 3,5 di Br PADAP system indicated that only single absorbing species exist under varied experimental conditions. Further the Job's method of continuous variations indicated a ratio of 1:2 for Th

: 3,5 di Br PADAP in the ternary species. As solvent extraction date indicated a 1:3 ratio for Th : TOPO, the final Th(IV)-TOPO 3,5 di Br PADAP ternary system is likely to be constitued with Th, TOPO and 3,5 di Br PADAP in the ratio 1:3:2.

9.2 Analytical Applications

Thorium is found to be associated with Mg, Ca, Fe, Al, Sn, Ti, Zr and U in nature. All these diverse elements interfere in the estimation of Th by many methods and its analysis in natural and synthetic materials is a challenging task. There is no single step practical method for separating thorium from all its associated elements. There are a few methods for its separation from Sn, Ti, Zr, Nb, and W

A combination of certain methods of separation will serve to separate thorium from virtually all other elements. There are many methods for the separation from titanium, zirconium and tungsten.

As thorium is found to be associated with many other substances in nature for example in monzite sand as phosphate with rare earths, Mg, Ca, Fe, Al, Sn, Ti, Zr and U, it needs prior separation from them and solvent extraction is a simple, rapid and effective technique for the separation of thorium.

The Th(IV)-TOPO-3,5 di Br PADAP method can tolerate large concentrations of complexing anions like, diethyldithiocarbamate, oxalate and phosphate, which can be used as masking agents or pre-extracting agents for the separation of matrix components.

It is established by analysing simulated synthetic samples that this method can be directly applied for the determination of Th (20-200 μg) in presence of moderately large concentrations of Fe(100 mg), Ca(2 mg), Al(7 mg), Mg(15 mg), Zn(2 mg), Cu(3 mg), Pb(2 mg), As(2 mg) and Mn(2 mg) and trace quantities of Ti, rare earths, Nb which are likely to exist as matrix and composition components.

However, for higher levels of matrix components a preseparation step is essential for the estimation of thorium by any method. As diethyldithio carbamate (DDTC) which forms quantitatively extractable chelates with many metal ions does not react with thorium, and as Th(IV)-TOPO-3,5 di Br PADAP method can tolerate large concentrations (100 mg) of DDTC, this situation has been exploited for the bulk separation of matrix and constituent components likely to be associated with thorium particularly in synthetic industrial materials like steels, high temperature alloys etc.

Since thorium often occurs in natural materials like rare earth elements, iron, titanium, etc. Its separation from them has been studied by many workers. This separation is easily carried out by solvent extraction because thorium(IV) forms more stable complexes than the rare earths and thus it is in general extracted better than the rare earths. However, there are many other elements that are co-extracted with thorium under the same conditions and the experimental conditions are to be critically optimised to make it selective.

Among oxygen containing solvents mesityl oxide is a very effective and selective solvent which extracts and separates thorium (IV) quantitatively from various associated matrix components when the aqueous phase is saturated with aluminium nitrate and acidified with nitric acid.

It is further observed that thorium in presence of large concentrations of titanium, iron, uranium and Zirconium can be effectively separated using aliquot 336 (a long chain tertiary amine) in 10 molar HCI medium into cyclohexane. Double extraction completely eliminates zirconium and titanium. The application of this method for the bulk separation of titanium and zirconium matrix has been tested in combination with mesityl oxide pre-concentration of thorium.

Thus the application of Th(IV)-TOPO-3,5 di Br PADAP method for the analysis of natural and industrial synthetic materials has been demonstrated by the author through systematic and detailed studies, and the best optimum conditions for the pre-extraction procedures for the effective separation of thorium are developed. These procedures can be directly applied for the analysis of complex materials of diversified matrix components either by :

(A) Bulk separation of matrix and constituent trace interfering elements from thorium by pre-extracting them with diethyl dithiocarbamate. This procedure can be rapidly and effectively applied for the analysis of steels, caper base alloys, magnesium base alloys, nickel base alloys.

(B) Selective pre-extraction of thorium with mesityl oxide from matrix and constituent components from 2M HNO_3. The method can be applied in the analysis of natural materials like monazite sand, ores, rocks etc., containing thorium along with rare earths, phosphate etc.

(C) Bulk extraction separation of titanium, iron, uranium and zirconium from Th with aliquot-336 into cyclohexane. This method can be applied in the analysis of natural materials like rutile, ilmenite samples and titanium base alloys.

The applicability of the optimised procedure for the bulk separation of matrix components by pre-extraction with DDTC into chloroform : acetone mixture has been established by analysing synthetic simulated samples of composition similar to steels, nickel, magnesium, copper base alloys containing Th(50-200 μg) along with matrix components like Al, MG, Ni, Cu, Fe and relatively low concentrations of trace element like Ti, Cr, Mn, Mo, Pb, rare earths and U. These investigations gave precise results of Th(IV) with standard deviation in the range 1.1-4.0 and coefficient of variation 1.1-4.9.

Applicability of the above procedure has also been demonstrated by analysing actual NBS steels and alloy samples NBS SRM 171, NBS SRM 867, NBS SRM 19g after spiking known amount of thorium (0.17%-0.2%) in a 0.5g of the corresponding steel samples.

The effective and selective pre-extraction of thorium with 4 methyl 1-3 pentene - 2 one (mesityl oxide) has been applied for the analysis of natural and synthetic industrial materials containing thorium by Th-TOPO-3,5 di Br PADAP method. The method is based on selective and quantitative extraction of thorium nitrate from solutions saturated with aluminium nitrate as saltingout agent with which equilibrium is attained rapidly. Thorium is then stripped from organic phase by shaking with water and estimated by Th-TOPO-3,5 di Br PADAP method. Uranium is the only element which eo-extracts and interferes which can be removed by pre-extraction with diethyl dithiocarbamate before mesityloxide extraction. Zirconium, vanadium, bismuth though co-extract into mesityloxide, they will not interfere, as in back extraction or in stripping with water they remain in the organic phase. The potentiality of this method is the separation of thorium from rare earths.

This method has been optimised such that Th can be effectively separated from large concentrations of Ni(II), Mg(II), Cu(II), Fe(III), and Al(III) as matrix components and low concentrations of Bi(III), V(V), Zr(IV), Co(II), Mn(II), Mg(II), Ca(II), Ti(IV), Nb(V) and rare earths as constituents, by selectively extracting it into mesityloxide at 2M HNO_3 medium in presence of saltingout agent like $Al(NO_3)_3$ 9 H_20.

This procedure has been tested by analysing synthetic simulated samples of composition similar to steels, nickel, magnesium, aluminium and copper base alloys in which thorium (40-200 μg) has been separated effectively from matrix components and estimated by Th-TOPO 3,5 di Br PADAP method which gave results with standard deviation (0.6-4.9) and coefficient of variation (0.6-3.2).

In the case of standard aluminium alloy NBS SRM 859, Th(0.18%) is spiked to the aluminium alloy solution and then analysed by the present procedure and values obtained by this method are reasonably accurate and precise when compared to spiked amounts.

However, for samples containing large concentrations of titanium, zirconium and uranium both the DDTC and mesityl oxide procedures do not give accurate results and need prior separation of these matrices from thorium for which a procedure has been optimised using aliquot-336.

The pre-extraction of matrix components like zirconium, titanium, uranium and iron using aliquot-336 from 10 M HCI media and then analysis of thorium levels using Th(IV)-TOPO-3,5 di Br PADAP method has been optimised such that complete procedure is directly applicable for such samples. Initially the applicability of the above procedure has been tested by analysing synthetic simulated samples containing zirconium, titanium, uranium, and iron as matrix components (100 mg) and thorium levels in the range (50-200 μg) which gave reasonably accurate results with standard deviation 1.5-4.2 and co-efficient of variation 2.0-3.2. The applicability of the above procedure has also been demonstrated by analysing accurately actual rutile, ilmenite samples after spiking 0.12%-0.14% of thorium which yielded results with standard deviation ranging from 1.9-2.8. For the analysis of titanium base alloy a known amount of thorium 0.2% is spiked to 0.5 gm of titanium base alloy NBS SRM 50(a) solution in 100 ml. and precisely analysed by Th(IV)-TOPO-3.5 di Br PADAP method.

Conclusions

(1) Th(IV) can be selectively extracted with TOPO as $Th(No_3)_4$ 3 TOPO in low nitric acid medium while in high nitric acid medium as an adduct [Th(IV) $(NO_3)_4$ 3 TOPO]nHNO_3 TOPO.

(2) The Th(IV)-TOPO-3,5 diBr PADAP method has dual advantages where TOPO selectively extracts thorium from high acid medium

when present along with diversified complex matrices and 3,5 diBr PADAP imparts necessary sensitivity to the colour reaction making it applicable from trace determination of thorium.

(3) This method can be directly applied for the precise analysis of Th(50-200 μg) even in presence of moderately large concentrations of Fe(II) (100 mg), Mg(II) (15 mg), Ca(II) (2 mg), Al(III) (7 mg), Mn(II) 2 mg), Zn(II) (2 mg), Cu(II) (2 mg) and Pb(II) (2 mg) without any pre-extraction of matrix.

(4) For higher concentration of matrix components this method can be applied after bulk separation of the matrix components by pre-extraction. The application of Th(IV)-TOPO-3,5 diBr PADAP method for the analysis of Th levels in natural and industrial synthetic materials with diversified complex matrices the following procedures (for the bulk separation of the matrix components by pre-extraction) are optimised to achieve and accurate results.

(A) Bulk separation of matrix and constituent elements from thorium by pre-extraction with diethyl dithiocarbamate. This procedure can be applied for the analysis of steels, copper base alloys, magnesium base alloys and nickel base alloys.

(B). Selective pre-extraction of thorium with mesityl oxide from 2 M HNO_3 with aluminium nitrate as salting out agent. The procedure can be applied for the analysis of thorium in natural materials like monazite sand and aluminum alloys.

(C) Bulk separation of Titanium, iron, uranium and zirconium, matrices by pre-extraction with aliquot-336. This method can be applied for the analysis of rutile, ilmenite samples and titanium base alloys.

These procedures can be applied either individually or in combination with another depending on the composition of matrix components of the sample.

(5) The Th(IV)-TOPO-3,5 diBr PADAP method is simple more rapid and precise. The co-efficient of variation varies between 0.8 and 2.4 for the estimation of thorium varying in the range 50-200 μg in the initial aqueous solutions.

This method gives more accurate and reproducible results of Th than the widely used arsenazo III methods which has less reproducibility as the sensitivity of the arsenazo method varies even with slight change in the acidity of the medium. This method is simple, rapid and involves clear separation by solvent extraction. Further the sensitivity is high enough for traces of thorium to be determined in various diversified complex matrices.

(6) The practical applicability of DDTC extraction method for the bulk separation of the matrix components has been demonstrated by analysing synthetic samples of simulated composition similar to Mg, Cu, Ni base alloys and steels with thorium in the range (0.05%-0.2%) in 0.5 gm of NBS steels and alloys sample solutions.

(7) The analytical capability of the optimised procedure for the selective separation and pre-concentration of thorium from matrix components like Mg, Al, Ni, Cu, Mn, Ca, Ce, La, Y, Pb and Zn with mesityl oxide and then precise estimation of thorium by Th(IV)-TOPO-3,5 diBr PADAP method has been established. This selective separation and pre-concentration of thorium from matrix components can be applied for the analysis of both natural and synthetic industrial materials.

The utility of this method has been demonstrated by analysing synthetic simulated samples of composition similar to steels, Mg, Al, Cu and Ni base alloys with Th levels (0.05%-0.2%(of the total matrix of 100 mg. components. This method can be advantageously applied for the analysis of Th in monazite sand. Spiked amounts of thorium (0.18%) in 0.5 gm of NBS SRM 859 aluminium alloy, solutions are also precisely analysed.

This method can be applied for the natural materials like monazite sand after precipitation of Th with oxalate.

(8) For the analysis of thorium in zirconium, titanium, uranium and iron matrix samples like rutile, ilmenite (as the DDTC and mesityl oxide procedures yielded incomplete separation) an efficient separation procedure by pre-extraction of the matrix components with aliquot-336 has been optimised such that Th(IV)-TOPO-3,5 diBr PADAP method can be adopted for samples of these matrices.

(9) The capability of this method, for the analysis of Th in monazite sand has been established by analysing actual monazite sand samples which yielded results with comparable with certified values.

(10) Based on extraction spectrophotometric data the mechanism of extraction of thorium with TOPO and formation of Th(IV)-TOPO-3,5 diBr PADAP ternary complex in non-aqueous medium has been explained.

TABLE

Comparison of Th(IV)-TOPO-3,5-diBr PADAP Method with Other Extractive Spectrophotometric Methods for the Determination of Thorium

S. No.	*Reagent*	*Conditions*	*Remarks*	*Ref.*
1.	Thoron I	0.3 M HCl medium molar absorptivity 1.7 x 10^4 l.mol.$^{-1}$ cm.$^{-1}$ λ max $^{450\,nm}$	The absorbance decreases with increasing acidity. The method is used for determination of Th in silicate rocks, ores.	22
2.	Arsenazo III	in strongly acidic medium λ max $^{655\,nm}$ molar absorptivity 1.15 x 10^5 l.mol.$^{-1}$cm.$^{-1}$	Acidity should be same in standard ards, blank and sample. The method is used for minerals, geological samples.	23
3.	1 (2 thiazolylazo) 2 naphthol (TAN)	Glycine buffer was used maintain pH 2.4-2.8 where the reaction was studied. λ max ^{555} E = 3.14 x 10^4 l.mol.$^{-1}$cm.$^{-1}$	Antipyrine was used for the enhancement of sensitivity	24
4.	4 (2 thiazolylazo) resorcinol (PAR)	In presence of 10% hexamethylene tetramine at pH 7.4 the reaction was studied λ max ^{615} E = 3.15 x 10^4 l.mol.$^{-1}$cm.$^{-1}$	Interference of Fe, EDTA, Cr, Cu, Ga studied	20
5.	4 (2 pyridylazo) resorcinol (PAR)	at pH 6.7-6.9 reaction was studied λ max $^{500\,nm}$	The interference of many metal ions observed	25
6.	TOPO-3.,5 diBr PADAP	extracted with TOPO from 6M nitric acid medium and reacted with 3,5-diBr PADAP in pyridine Tolune medium λ max $^{575\,nm}$ E = 1.2 x 10^5 l.mol.$^{-1}$cm.$^{-1}$	Highly selective and sensitive and rapid	Present method

References

1. Honigschmidt O., *Monatsh* 27, 205, 1906.
2. Honigschmidt O., *Compt. Rend* 142, 157, 1906.
3. Honigschmidt O., *Compt. Rend* 142, 280, 1906.
4. Filyand M.A., Semenova E.I., *Properties of Rare Elements Hand Book.*
5. Grube G., *Metallurgizdt.* 1953.
6. Chauvenet E. *Ann. Chim.* (8) 23, 425, 1911.
7. Leoxtis T.E., *J. Metals* 4, 287, 1952.
8. Mayer A., Bradshaw G., *Analyst* 77, 154, 1952.
9. Nelson K.C., Strieter F.P. *Trans. Am. Foundrymen's Assoc.* 58, 400, 1950.
10. Kettembeil *Z. Anorg. Chem.* 38, 217, 1937.
11. Parks W.G., Prime G.E., *J. Am. Chem. Soc.*, 58, 1413, 1936.
12. Shibata S., *Cheletes in analytical chemistry*, Vol. 4, Felschka H.A., Barnard A.J. Ed., Marcel Dekker, New York, 1, 1972.
13. Anderson G., Nickless B., *Analyst* 92, 132, 1962.
14. S.I. Gusev and L.M. Schurova *Zh. Analit. Khim.*, 21, 1042, 1966.
15. Goto K., *Anal. Chem.* 38, 493, 1966.
16. Bryan L.M. *Anal. Chem.* 39, 1706, 1967.
17. Koichi Tonosaki, Takayoshi Sakar., *Bull. Chem. Soc.*, Japan.4(2), 456, 1969.
18. Malik W.U.J. *Less. Common Metals* 16, 385, 1968.
19. Adam J., Pribil R., *Talanta* 16, 1596, 1969.
20. Koichi Tonosaki, Takayoshi Sakai, *Bull. Chem. Soc.*, Japan 42(9), 2719, 1969.
21. Ramakrishna T.V., *Talanta* 26, 499, 1979.
22. Palmer A.R. *Anal. Chim. Acta*, 14, 162, 1956.
23. Savvin S.B. *Talanta* 8, 673, 1961.
24. Sharma C.D., Eshwar M.C., *J. Radio and Nucl. Chem.* 91(2), 323, 1985.
25. Vibhute C.P. Khopkar S.M., *J. Radio and Nucl. Chem.*, 97(1), 3, 1986.